# Water Rock Interaction in Underground Coal Mining

Cun Zhang

# Water Rock Interaction in Underground Coal Mining

## Problems and Analysis

Cun Zhang
School of Energy and Mining Engineering
China University of Mining and Technology (Beijing)
Beijing, China

ISBN 978-981-95-9956-1 ISBN 978-981-95-9957-8 (eBook)
https://doi.org/10.1007/978-981-95-9957-8

This work was supported by National Key R&D Program of China (2023YFC3012101), Ordos Youth Talent Technology Project (RC20250108) and National Natural Science Foundation of China (52474161).



This Springer imprint is published by the registered company Springer Nature Singapore Pte Ltd.
The registered company address is: 152 Beach Road, #21-01/04 Gateway East, Singapore 189721, Singapore

# Preface

The water-rock interaction (WRI) is one of the research focuses in the field of geology and geotechnical engineering. Many physical and chemical reactions are involved in the process of WRI, mainly including lubrication, precipitation, oxidation-reduction, ion-exchange, etc. Many geological hazards and engineering safety problems relate to WRI, e.g., slope stability, reservoir dam stability, rock bursting, karst collapses, and water inrush. The influence of water on the mechanical and seepage behavior of rock mass is the basis for the analysis and interpretation of the problems mentioned above. In recent years, a large body of research has investigated the effect of water on rock mechanical characteristics. It is generally found that the presence of water weakens the mechanical parameters (e.g., elastic modulus, compressive strength, cohesion, and tensile strength) as well as the brittle property of a rock, and also changes the fragmentation distribution after rock failure. On the one hand, the existence of water increases the pore pressure in soil and rock mass, which reduces the effective stress of skeleton particles in the medium, so as to change the physical and mechanical parameters of the media. On the other hand, the water may dissolve mineral compositions and cement between grains, so as to produce new mineral compositions. If the water contains some corrosive mineral components, the WRI would be more serious. For example, rocks with higher clay mineral contents are more susceptible to water. There are many aquifers in coal measure strata, and coal mining activities disturb these aquifers. Therefore, it is inevitable that water would flow into mined areas, threatening mining safety and even causing water inrush accidents. In addition, underground rock structures are further destabilized under the softening effect of water, which might result in safety accidents. Mine water disaster is regarded as one of the most critical problems in coal mining, and thus, has been a research hotspot in coal mining. The WRI process plays a significant role in water-related disasters and is the key to uncover the mechanisms underlying these disasters. The objective of this book is to review and study the WRI in coal mining environments.

This book is divided into seven chapters. Within those seven chapters, Chap. 1 gives a literature review on the WRI problems that may occur in coal mining environments. For each problem, the most important characteristics of WRI are illustrated. Laboratory analysis methods and relevant measures of WRI are presented, and primary conclusions on problems of (1) water weakening of intact rock, (2) stress-flow characteristics of fracture rock mass, and (3) compaction-flow characteristics of broken rock mass are outlined. Simulation methods of WRI in a coal mining environment are summarized in Chap. 1, which can be generally divided into continuous rock mass simulation method and broken rock mass simulation method. Their capabilities and limitations are also discussed. Finally, concluding remarks and prospects are presented in Chap. 1.

Chapter 2 focuses on the water weakening of rock strength and its influencing factors (water content, immersion time, and wetting–drying cycles). The strength of the rock mass decreases to varying degrees with water content, immersion time, and wetting–drying cycles depending on the rock mass type and mineral composition. The corresponding acoustic emission count and intensity and infrared radiation intensity also weaken accordingly. WRI enhances the plasticity of rock mass and reduces its brittleness. Various microscopic methods for studying the pore characterization and weakening mechanism of the WRI are compared and analyzed. Various methods should be adopted to study the pore evolution of WRI comprehensively. Microscopic methods are used to study the weakening mechanism of WRI. In future work, the mechanical parameters of rocks weakened under long-term water immersion (over years) should be considered, and more attention should be paid to how the laboratory scale is applied to the engineering scale.

Chapter 3 proposes a method to determine the threshold value of various mineral components and pores inside the coal sample through computed tomography (CT) scanning, X-ray diffractometer (XRD) test, and nuclear magnetic resonance (NMR) technology. This method determines the porosity of the coal by NMR and then determines the pore reconstruction threshold. The type and composition of coal sample minerals were determined by XRD. Combined with the difference in the density of each mineral, the classification threshold of each type of mineral can be determined. We first determined the representative elementary volume through pore and mineral reconstruction of coal with different sizes. On this basis, we selected three cubes with a side length of 300 pixels and different positions for pore and mineral reconstruction. The extraction of different mineral components is helpful for the research on the evolution of various minerals before and after water immersion.

In Chap. 4, the evolution of pore structure and mechanical behavior of sandstones at different immersion times (0d, 3d, 30d, 90d, 180d, and 360d) are studied. A series of physico-chemical experiments and uniaxial compression tests (with simultaneous acquisition of acoustic emission (AE) signals) are conducted to explore the weakening mechanism and failure mode of sandstones during long-term immersion. Variation of sandstone pore structure during long-term immersion is revealed using multi-parameter. The weakening mechanism of WRI under long-term immersion conditions is explained. The study results enhance our understanding of weakening mechanisms at long-term immersion and provide a basis for assessing the long-term stability of underground engineering.

Chapter 5 proposes the simulation methods of fluid-solid coupling for water immersion weakening in coal mining. The permeability and strength of coal-rock mass can be realized to update real-time during mining and water immersing in the methods. The implementation process of the simulation is discussed with the discrete element method and finite element method. The permeability is updated through stress and crack opening in the discrete element simulation method. Meanwhile, the variable joint stiffness simulation method that considers changes in crack compressibility is proposed. When the water seeps into the joint crack, the update of the water content is calculated by the change in joint opening. Before the coal-rock mass is yielded, the weakening strength is updated based on the relationship between water content and coal-rock mass strength. Compared with the discrete element simulation method, the update of the permeability is needed to consider anisotropy in the finite element simulation method. The update of the saturated water content is calculated based on the porosity. Finally, the fluid-solid coupling simulation method of water immersion weakening proposed is discussed and prospected. It is considered to be more helpful for the analysis of WRI engineering problems in coal mine, that mastering the scale effect of mechanical seepage characteristics for coal-rock mass and implementing long-term water immersion weakening in simulation.

Chapter 6 simulates the failure of residual coal pillar under the coupling of mining stress and water immersion based on fluid-solid coupling methods in Chap. 5. A quantitative index of the damage ratio of coal pillars was proposed. The numerical parameters are calibrated with measurement data from coal pillar stress. In the simulation process, the evolution of the strength, damage ratio, water pressure, stress, strain, and other parameters in coal pillars under the effect of mining-water invasion is realized, and the progressive failure mechanisms of coal pillars under mining-water immersion are revealed. On this basis, the effects of goaf water-level height and coal-pillar width on coal-pillar stability are quantitatively analyzed, and the critical goaf water-level height corresponding to different coal pillar widths is obtained.

Chapter 7 simulates surrounding rock weakening in water-filled fracture structure roadway based on fluid-solid coupling methods in Chap. 5. A damage quantitative index suitable for numerical simulation are proposed. Based on the quantitative relationship of water immersion weakening of rock samples in laboratory tests, the mechanical properties of the coal and rock mass with different water immersion were calibrated. The progressive failure characteristics of the roadway surrounding rock in a fault structure area with water-rich roofs were analyzed. According to the progressive failure characteristics of the RSR from water-rich roofs, a control strategy of dewatering the roof aquifer near the fault and strengthening the roadway support is proposed.

The publication of this book is financially supported by the National Key R&D Program of China (2023YFC3012101), the Ordos Youth Talent Technology Project (RC20250108), the National Natural Science Foundation of China (52474161), the Postdoctoral Research Foundation of China (2025T180509), and the Fundamental Research Funds for the Central Universities (2024ZKPYNY01). The authors also would like to thank the support provided by the China University of Mining and Technology (Beijing), Inner Mongolia Research Institute, China University of Mining and Technology (Beijing), and Anhui University of Science and Technology.

Beijing, China
January 2026

Cun Zhang

**Competing Interests** The author has no competing interests to declare that are relevant to the content of this manuscript.

# Abstract

Underground coal mining has a strong disturbance on surrounding rock strata, causing inevitable water inflow into mined spaces, which seriously affects safety production. Mine water disaster is regarded as one of the most critical problems in coal mining, and thus, has been a research hotspot in coal mining. This book focuses on water-rock interaction (WRI) in underground coal mining and lists the main issues of WRI. The corresponding analysis methods (laboratory testing and numerical simulation) of WRI in coal mining are discussed. The research related to WRI in underground coal mining mainly focuses on the water inrush form roof and floor, water weakening of roadway surrounding rock (coal pillar), water inrush from fault or collapse column, and the utilization of abandoned mines. Laboratory tests (on intact rock mass, fractured rock mass, and broken rock mass) approaches numerical simulation methods of WRI are summarized, and their advantages and limitations are discussed. Based on the WRI problems and analysis in underground coal mining, future works on laboratory testing and numerical simulation is prospected and studied.

Aiming at solving this issue, this book summarizes the recent research work regarding the WRI in underground coal mining by the authors. This book is divided into seven chapters. Within those seven chapters, the WRI problems that may occur in coal mining environments are summarized, strength weakening of the rock and its micromechanism in WRI are analyzed, an accurate characterization method of pores and various minerals is proposed, the mechanical characteristics and damage mechanism of sandstone under long-term immersion are studied, the simulation methods of fluid-solid coupling for water immersion weakening are proposed, the feasibility of the simulation method is verified by simulating surrounding rock weakening in water-filled fracture structure roadway, and the failure of residual coal pillar under the coupling of mining stress and water immersion is analyzed.

This book can provide guidance, experience, and reference for scientists and engineers who work in mining engineering, civil engineering, and geotechnical engineering.

**Keywords** Water-rock interaction; Underground coal mining; Weakening mechanism; 3D reconstruction; Long-term immersion; Fluid-solid coupling simulation; Coal pillar failure; Roadway failure

# Contents


**1 Introduction** ........ 1
- 1.1 Water-Rock Interaction Problems in Coal Mining ........ 1
  - 1.1.1 Stability of Roadway in a Water Environment ........ 1
  - 1.1.2 Mine Water Inrush Through Geological Anomalies ........ 2
  - 1.1.3 Water Seepage from Surrounding Measures Under Mining Disturbances ........ 3
  - 1.1.4 Underground Reservoir and Abandoned Flooded Mine ........ 5
- 1.2 Laboratory Analysis of WRI in a Coal Mining Environment ........ 7
  - 1.2.1 Water Wakening of Intact Rock ........ 8
  - 1.2.2 Stress-Flow Characteristics of Fracture Rock Mass ........ 9
  - 1.2.3 Compaction-Flow Characteristics of Broken Rock Mass ........ 11
- 1.3 Simulation of WRI in a Coal Mining Environment ........ 13
  - 1.3.1 Continuous Rock Mass Simulation ........ 13
  - 1.3.2 Broken Rock Mass Simulation ........ 18
- 1.4 Concluding Remarks and Prospects ........ 20

**2 Strength Weakening and Its Micromechanism in Water–Rock Interaction** ........ 25
- 2.1 Introduction ........ 25
- 2.2 Water Weakening Characteristics and Its Influencing Factors ........ 26
  - 2.2.1 Water Content ........ 27
  - 2.2.2 Wetting-Drying Cycles ........ 31
  - 2.2.3 Immersion Time ........ 31
  - 2.2.4 Other WRI-Related Factors ........ 35
- 2.3 Weakening Mechanism and Microscopic Characterization ........ 35
- 2.4 Conclusion and Prospect ........ 40

**3 Accurate Characterization Method of Pores and Various Minerals** ........ 43
- 3.1 Introduction ........ 43
- 3.2 Test Samples and CT Scans ........ 44
  - 3.2.1 Test Sample ........ 44
  - 3.2.2 CT Scan ........ 47

3.3 Threshold Segmentation for Pores and Minerals . . . 48
3.3.1 Model 3D Visualization . . . 48
3.3.2 Select the Analysis Area . . . 49
3.3.3 Threshold Segmentation . . . 49
3.4 3D Reconstruction for Pores and Minerals . . . 52
3.4.1 Division of Pores . . . 52
3.4.2 Division of Mineral Components . . . 54
3.5 Discussion and Conclusion . . . 57

**4 Mechanical Characteristics and Damage Mechanism of Sandstone Under Long-Term Immersion** . . . 61
4.1 Introduction . . . 61
4.2 Experimental Method . . . 63
4.2.1 Preparation of Rock Samples . . . 63
4.2.2 Testing Equipment and Scheme . . . 65
4.3 Physical and Mechanical Properties of Sandstone with Different Soaking Time . . . 66
4.3.1 Analysis of P-Wave Velocity . . . 66
4.3.2 Microstructural Descriptions . . . 67
4.3.3 Mechanical Properties and AE Responses . . . 70
4.3.4 Mineral Dissolution . . . 78
4.4 Long-Term Water Immersion Weakening Mechanism . . . 79
4.5 Conclusions . . . 83

**5 Fluid-Solid Coupling Simulation Method for Water Immersion Weakening in Coal Mining** . . . 85
5.1 Introduction . . . 85
5.2 Fluid-Solid Coupling Softening Method in Finite Element Analysis . . . 86
5.2.1 Permeability Calculation Model . . . 87
5.2.2 Water Invasion Weakening . . . 88
5.3 Fluid-Solid Coupling Softening Method in Discrete Element Analysis . . . 91
5.3.1 Permeability Calculation . . . 92
5.3.2 Water Invasion Weakening Criteria . . . 93
5.3.3 Weakening Coefficient Acquisition . . . 94
5.3.4 Damage Ratio Calculation . . . 96
5.4 Discussion and Outlook . . . 96

**6 Failure Analysis of Residual Coal Pillar Under the Coupling of Mining Stress and Water Immersion** . . . 99
6.1 Introduction . . . 99
6.2 Process of Pillar and Gate Road Failure . . . 101
6.2.1 Engineering Geological Conditions . . . 101
6.2.2 Instability Analysis of Gate Road and Coal Pillar . . . 103

6.3 Numerical Simulation of the Stability of Coal Pillars . . . . . . . . . . . 106
6.3.1 Numerical Model Construction . . . . . . . . . . . . . . . . . . . . . . . 106
6.3.2 Mining Stress Inversion of Coal Pillars . . . . . . . . . . . . . . . . 109
6.3.3 Stability Analysis of Coal-Pillar Mining-Water Invasion Weakening . . . . . . . . . . . . . . . . . . . . . . . . . . . . . . . 109
6.4 Stability Control Measures for Coal Pillars . . . . . . . . . . . . . . . . . . 115
6.4.1 Suitable Pillar Size . . . . . . . . . . . . . . . . . . . . . . . . . . . . . . . 115
6.4.2 Suitable Height of the Goaf-Water Level . . . . . . . . . . . . . . . 117
6.4.3 Field Practice . . . . . . . . . . . . . . . . . . . . . . . . . . . . . . . . . . . 122
6.5 Conclusions . . . . . . . . . . . . . . . . . . . . . . . . . . . . . . . . . . . . . . . 124

**7 Progressive Failure of Roadways in a Fault Structure Area with Water-Rich Roofs** . . . . . . . . . . . . . . . . . . . . . . . . . . . . . . . . 127
7.1 Introduction . . . . . . . . . . . . . . . . . . . . . . . . . . . . . . . . . . . . . . . 127
7.2 Field Measurement of Surrounding Rock Failure of Roadway . . . . 128
7.2.1 Engineering Geological Condition . . . . . . . . . . . . . . . . . . . 128
7.2.2 Instability Characteristics of Roadway Surrounding Rock . . . 130
7.3 Numerical Model and Parameters Determination in Engineering . . . 132
7.3.1 Numerical Model Construction . . . . . . . . . . . . . . . . . . . . . . 132
7.3.2 Parameter Calibrations . . . . . . . . . . . . . . . . . . . . . . . . . . . . 134
7.4 Failure Analysis of Water Invasion Weakening of Roadway Surrounding Rock . . . . . . . . . . . . . . . . . . . . . . . . . . . . . . . . . . . . 136
7.5 Control Techniques . . . . . . . . . . . . . . . . . . . . . . . . . . . . . . . . . . 142
7.6 Conclusions . . . . . . . . . . . . . . . . . . . . . . . . . . . . . . . . . . . . . . . 145

**8 Conclusion** . . . . . . . . . . . . . . . . . . . . . . . . . . . . . . . . . . . . . . . . 147
8.1 The Integrated WRI Analysis Framework . . . . . . . . . . . . . . . . . . 147
8.2 Synergistic Contributions of the Framework . . . . . . . . . . . . . . . . . 148
8.3 Future Outlook . . . . . . . . . . . . . . . . . . . . . . . . . . . . . . . . . . . . . . 149

**References** . . . . . . . . . . . . . . . . . . . . . . . . . . . . . . . . . . . . . . . . . 151

# Chapter 1
# Introduction

## 1.1 Water-Rock Interaction Problems in Coal Mining

During coal mining, the research on WRI mainly focuses on water inrush from the roof and floor (Han et al., 2022b; Qin et al., 2020), stability of roadway (coal pillar) in water-rich environments (Han et al., 2022a; Li & Liang, 2009), water inrush from fault or collapse column (Ma et al., 2021; Wang et al., 2020a; Yao et al., 2018), and utilization of abandoned mines (Bian et al., 2021; Gu et al., 2021; Li et al., 2021d; Zhang et al., 2021c). With the increase in research on mine underground reservoir technology and the reuse of abandoned mine resources, the research on the rock characteristics of water immersion weakening (broken rock mass and residual coal pillar in the goaf) has gradually become a hot spot (Zhang et al., 2021a, 2021b).

### *1.1.1 Stability of Roadway in a Water Environment*

The adsorption of groundwater would soften the surrounding rock strength of a roadway. Under this condition, the surrounding rock of a roadway is prone to experience nonlinear deformation and severe failures, such as roof subsidence, spalling rib, and floor heave. Therefore, water, as a key factor affecting the behavior of the rock mass, has become a study hotspot on the stability of roadway, as shown in Fig. 1.1. To characterize the water absorption and softening of the roadway surrounding rock mass, numerous laboratory tests have been carried out (Liu et al., 2012; Xiong et al., 2011; Yang et al., 2017; Zhang et al., 2021g). He et al. (2008) researched the effect of water absorption parameters on the mechanical properties of mudstone by using the deep soft rock hydraulic experiment system. X-ray diffractometer (XRD) and scanning electron microscope (SEM) experiments are commonly used to observe the changes in the micropore structures and mineral composition in the tests. Yang et al. (2006) obtained the changes in mineral particles


C. Zhang, *Water Rock Interaction in Underground Coal Mining*,
https://doi.org/10.1007/978-981-95-9957-8_1

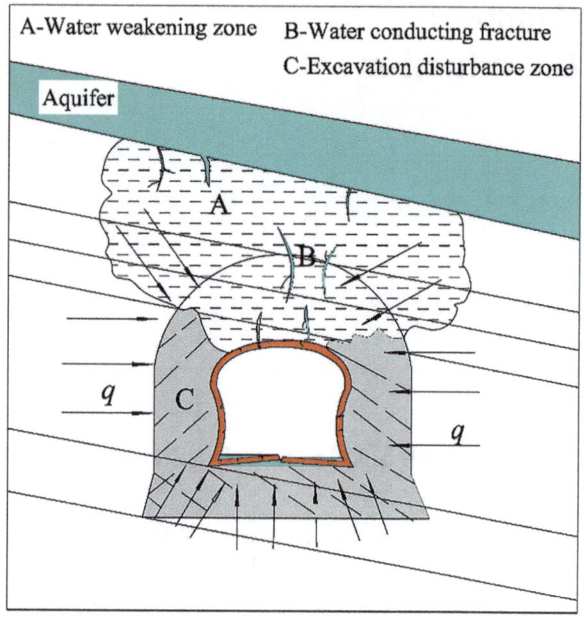


**Fig. 1.1** Roadway surrounding rock weakening and instability with water immersion

and cements between dry and saturated slate rocks. Li et al. (2011) analyzed mineral compositions of mudstone from different mining areas in China and related the pore structure to water absorption parameters. In many cases, the WRI not only weakens the rock strength, but also corrodes the reinforcement elements (bolt and anchor cable), which degrades the reinforcements and makes the roadway unstable (Showkati et al., 2021). Thus, there are two major concerns in the WRI analysis: the influence of (1) the water absorption (water content) and (2) the water immersion time on the strength and mechanical behavior of the surrounding rock mass.

### *1.1.2 Mine Water Inrush Through Geological Anomalies*

Mine water inrush is a geological disaster in the process of coal mining. The inrush disaster depends on several critical factors, i.e., water sources, water flow channels, water volume, and water pressure. Moreover, most water inrush occurs in abnormal geological bodies (e.g., collapse columns and faults), which provide the primary channels for water inflow (Fig. 1.2). As a special geological structure with good permeability and poor compactness, rock strata near collapse columns and faults are mostly debris. These geological structures are favorable channels to cause water inrush, especially under the conditions of high water pressure and strong mining disturbances.

With the long-term WRI, the internal filling particles in collapse columns begin to lose; this increases the porosity and permeability in the collapse column. The

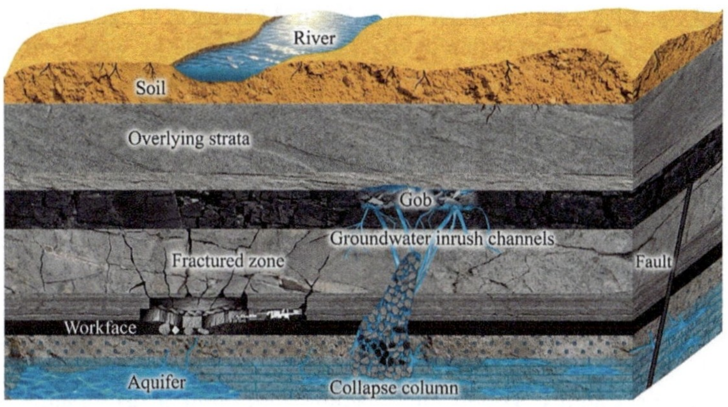


**Fig. 1.2** Schematic diagram of complex geological features near coal seams

filling material inside the collapse column will be activated, inducing the mine water inrush (Zhang et al., 2016a, 2021e; Gu et al., 2021). In addition, when mining activities are carried out above an aquifer, the karst collapse columns and the surrounding rock floor are prone to experience compression-tension-compression cyclic loads (Bai & Tu, 2019), which continuously damage the floor rock mass. Once the broken rock mass (such as argillaceous and siliceous) in the collapse column is insufficient to bear the mining-induced stress and/or hydraulic erosion conditions, mine water inrush would occur. In addition to the collapse column, geologic discontinuities (e.g., faults) are also common channels for mine water inrush. Under the action of water erosion, clay minerals may peel off from the fault skeleton into the fluid and flow out, which increases the permeability and porosity of the rock mass, as well as the fluid velocity (Cao et al., 2022; Wang et al., 2014).

WRI in collapse columns and faults belongs to variable mass seepage. The phenomenon of variable mass seepage generally exists in all kinds of rock and soil media, which is very harmful to engineering safety. Therefore, the WRI in the process of water inrush from faults and collapse columns is mainly the variable mass seepage of broken rock mass. Critical factors in this process usually include water pressure, flow velocity, external stress environment, and porosity of broken rock mass.

### *1.1.3 Water Seepage from Surrounding Measures Under Mining Disturbances*

Coal mining will cause significant disturbances to the roof and floor strata. Generally, there are caving, fracturing, and bending subsidence zones in the roof. In the floor, there can also be three zones, i.e., the confined water lifting zone, floor fracturing zone, and water barrier zone, as shown in Fig. 1.3 (Zhang et al., 2019a). Therefore, once the aquifer is located within the fracture zone, the aquifer water will flow into the longwall face, which might have a certain impact on safe production. For roof aquifers, especially shallow buried coal seams (i.e., when mining-induced fractures reach the surface), the damaged aquifer directly affects the surface ecology. Under the influence of mining, the disturbance scope in the seepage field is usually larger than that of the stress field, mainly due to the water supply (Cheng et al., 2020). The change in permeability varies in different zones near the longwall panel, and the increase in horizontal permeability generally exceeds that of vertical permeability, as shown in Fig. 1.4 (Zhang et al., 2020a). In the caved zone and fractured zone, permeability in horizontal and vertical directions will rise sharply, while in the constrained zone only the horizontal permeability is likely to increase, with little change in vertical permeability. For the floor aquifer, the mining disturbance is much smaller than that of the roof. However, the floor strata sometimes include aquifers with high water pressure (i.e., confined water in Fig. 1.3). The confined water lifting zone develops during mining (Liu et al., 2022). With the mining of the coal seam, floor fractures may connect with the confined water lifting zone; as a consequence, floor water inrush will occur. Therefore, water inrush from the roof and floor involves stress-fracture-seepage coupling in the rock mass.

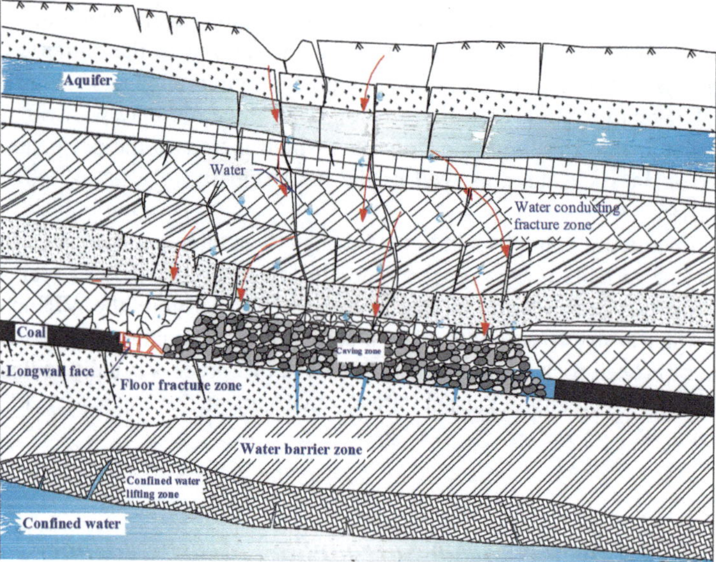


**Fig. 1.3** Representative fracture distributions near a longwall mining panel

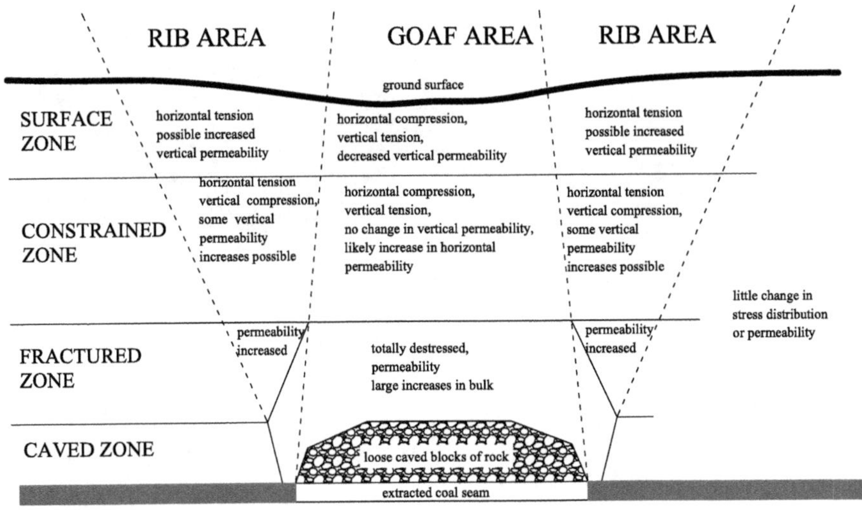


**Fig. 1.4** A hydrogeological model after coal mining

## *1.1.4 Underground Reservoir and Abandoned Flooded Mine*

Using the abandoned mine goaf to build an underground reservoir has been proposed in recent years, and the underground reservoir can be employed to construct pumped storage power stations. These images not only make full use of the abandoned underground spaces, but also promote environmental and ecological protection (Fan et al., 2020; Menéndez et al., 2019; Pujades et al., 2016; Zhang et al., 2021b). Unlike the existing underground reservoir, which only uses relatively stable shafts and roadways as the water storage space, the caving zone-based reservoirs provide much larger water storage spaces (Fang et al., 2019; Pang et al., 2019; Zhang et al., 2021a). However, the goaf is significantly affected by mining activities and the surrounding rock mass behavior. Changes in mechanical properties in the process of water storage will directly affect the stability of the goaf reservoir.

Coal pillar dam, residual coal pillar, and broken coal and rock mass (as displayed in Fig. 1.5) are the main bearing structures in the goaf reservoir. The micropores and microfractures in these bearing structures are complex and affected by dynamic and static loads, long-term water immersion, and circulation of water storage and drainage. Instability mechanism and seepage capacity of these bearing structures have become the core problems for safe and efficient operation of coal mine underground reservoir. Thus, the coupling effects of stress field and seepage field lead to a "damage - seepage - cumulative damage" failure process of the coal pillar dam. At present, the progressive failure characteristics of coal pillar dam under the coupling effect of mining and water invasion are primarily investigated in laboratory-scale tests, e.g., the water contents effect (Yao et al., 2019a, 2020a), water immersion time effect (Sadeghiamirshahidi & Vitton, 2019), seepage water pressure effect (Li

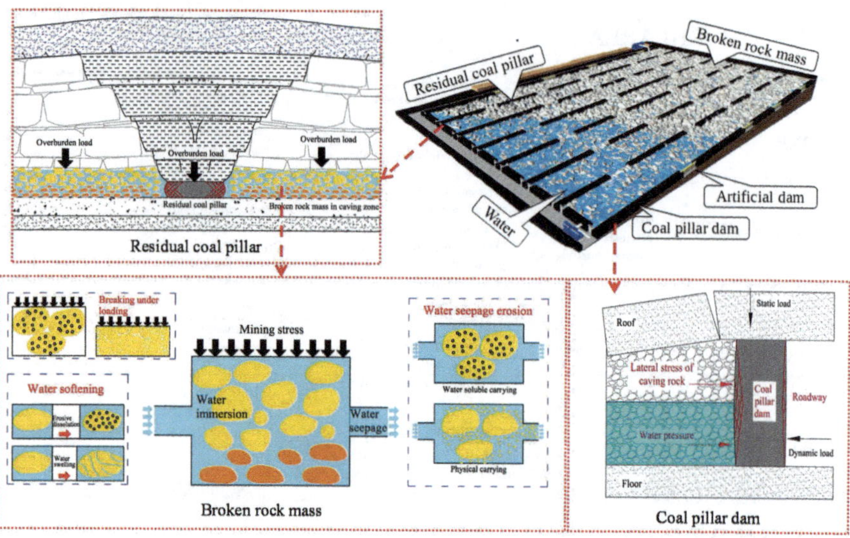


**Fig. 1.5** Schematic diagram of the primary bearing structures of the underground reservoirs

et al., 2020b), dry-wet cycles effect (Liu et al., 2020), and loading pattern effect on the strength of laboratory-scale coal specimens (Zhong et al., 2019).

The chain pillars between adjacent longwall panels transform into residual coal pillars under the action of dynamic and static loads (Fig. 1.5). When the underground reservoir starts to store water, the damage to the residual coal pillar accumulates because the residual coal pillar has not been reinforced. According to previous research, the residual coal pillars still play a key bearing role in controlling the subsidence of the roof. However, the bearing capacity of the residual coal pillars degrades under the weakening effect of long-term immersion, which may eventually lead to the instability of the bearing systems (i.e., residual coal pillars and coal pillar dam).

In the goaf reservoir, the broken rocks (i.e., caved roofs) in the goaf will be in the water immersion and loading environment for a long time. Under the WRI, the internal structures (e.g., microporous, microcracks, and mineral components) of the broken rock vary. The pore structure changes directly affect the water storage capacity of the reservoir. The WRI of broken rock mainly includes water immersion softening and seepage erosion, as shown in Fig. 1.5. The water softening effects usually contain the influence of water immersion time or water content on the strength of broken rock. The influence of water seepage is similar to the variable mass seepage in faults and collapse columns.

In addition, abandoned coal mines usually no longer discharge water, which leads to some goaves filled with water. The WRI between mine water and residual coal pillars will further weaken the strength of these pillars, resulting in coal pillar collapse (Andrews et al., 2020; Sadeghiamirshahidi & Vitton, 2019). In some countries, the rise of the water level after the closure of the coal mine causes the rise of the surface (Dudek et al., 2020; Zhao & Konietzky, 2020). Similarly, the construction of mine underground reservoirs and pumped storage power stations also impacts underground water quality due to the chemical interaction between rock and water (Fan et al., 2020).

## 1.2 Laboratory Analysis of WRI in a Coal Mining Environment

Based on the WRI conditions, many attempts have been made to mimic field environments in laboratory tests, as summarized in Table 1.1. The main objects of laboratory WRI analysis can be divided into intact rock mass (usually processed into a standard cylinder, disk, or cube specimens in the laboratory), fractured rock mass (mainly single fracture rock mass), and broken rock mass. The engineering problems of WRI primarily focus on the instability of surrounding rock and coal pillars, water inrush, and pore change caused by softening, seepage, and erosion. In laboratory testing, strength weakening due to moisture content, water immersion time, and wetting-drying cycles is commonly considered. In laboratory flow testing, the influence of water pressure and flow rate can be involved.

**Table 1.1** Methods to mimic field WRI in laboratory analysis

| Main research objects | Engineering problems | Water parameters | Laboratory testing |
|---|---|---|---|
| Intact rock | Roadway surrounding rock softens with water | Immersion time, moisture content, wetting-drying cycles | Tensile, compression, shear, and impact tests |
| | Residual coal pillar/coal pillar dam damage | Immersion time, water pressure, circulating water immersion | Tensile, compression, shear, and impact tests |
| Fractured rock mass | Water inrush from roof and floor | Water pressure/seepage velocity | Water seepage test |
| | Water seepage of residual coal pillar/coal pillar dam | Water pressure/seepage velocity | Water seepage test |
| Broken rock mass | Water inrush from fault and collapse column | Water pressure/seepage velocity | Water seepage test |
| | Compaction characteristics of caving zone | Immersion time | Compression tests |
| | Water storage capacity (porosity) of caving zone | Water pressure/seepage velocity, circulating water flow | Water seepage compaction test |

### 1.2.1 Water Wakening of Intact Rock

Mechanical experiments combined with the corresponding characterization means are commonly used to study the weakening characteristics of WRI. In conventional mechanical experiments (uniaxial and triaxial compression test, Brazilian splitting test, shear test, point load test, needle penetration test, Hopkins impact test, as shown in Fig. 1.6), we can obtain the weakening parameters of shear strength, tensile strength, compressive strength, hardness, and dynamic load strength. In the tests, in addition to obtaining the stress-strain curve, acoustic emission (AE), infrared radiation temperature (IRT), Digital Image Correlation (DIC), 3D laser scanning, and other methods can help characterize the influence of WRI (Fig. 1.6).

The weakening analysis of WRI mainly focuses on the influence of water content, immersion time, and wetting-drying (WD) cycles on the rock strengths, as well as the corresponding AE and infrared radiation (IR) characteristics. The strengths of rock mass decrease in varying degrees with water content, immersion time, and WD cycles, which are related to the type of rock mass and mineral composition (Chen et al., 2019b; Li et al., 2019a; Liu et al., 2019; Ranjith et al., 2008; Zhu et al., 2020a). Generally, the strengths decrease exponentially with the increase in water content (Hashiba & Fukui, 2015; Li et al., 2021a; Liu et al., 2019; Lu et al., 2021;

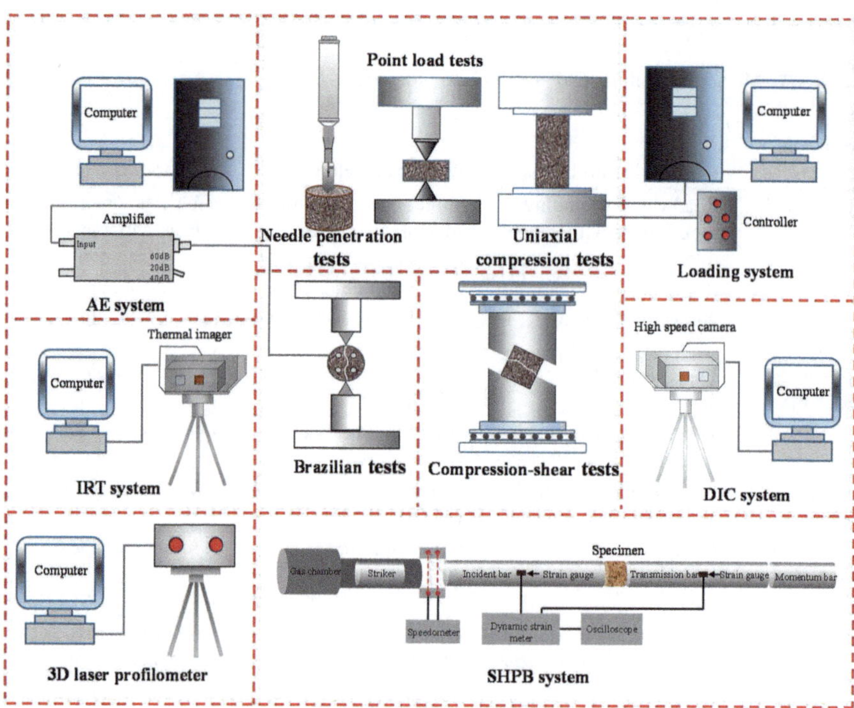


**Fig. 1.6** WRI mechanical tests and the primary analysis means

Luo, 2020). Some quantitative relationships (fitting formula) between mechanical parameters and water content are obtained by laboratory tests (Azhar et al., 2020; Chen et al., 2021a; Liu et al., 2021; Rabat et al., 2020; Tang, 2018). The dynamic strength and dissipated energy decrease, but the elastic modulus increases with the water content (Cai et al., 2020b; Yao et al., 2020b; Zhu et al., 2021).

Previous laboratory tests show that WD cycle treatment has a significant impact on the mechanical properties of rock (Chen et al., 2019c; Liu & Zhang, 2020; Momeni et al., 2017; Zhou et al., 2018a; He et al., 2020a; Huang et al., 2022). With the WD cycle increasing, porosity and water absorption present monotonically increase, while other strength parameters (shear, tension, and compression) generally decrease (Gratchev et al., 2019; Sun & Zhang, 2019; Khanlari & Abdilor, 2015; Huang et al., 2010; Zhao et al., 2018a). However, it is also found that the decrease of mechanical and physical parameters gradually diminishes as the WD cycle increases (Cai et al., 2020a; Fu et al., 2017; Li et al., 2021c; Wu et al., 2020b; Yao et al., 2019b).

In underground coal mining, roof and/or floor strata can be in the water immersion environment for months or even years; it is found that the immersion time has a significant effect on rock strength (Bai et al., 2016a). As the immersion time increases, the mechanical parameters and brittleness of the rock obviously weaken, and the failure mode transforms from unstable failure to stable failure. However, the roughness of the fracture planes after failure usually increases (Ai et al., 2021; Ma et al., 2021; Noël et al., 2021). Compared with dry conditions, water weakening tests usually obtain less AE counts and weak infrared radiation intensity (Guo et al., 2018; Li et al., 2021b; Lin et al., 2019; Sun et al., 2021b; Zhou et al., 2018c). In general, WRI enhances the plasticity of coal and rock mass and reduces the brittleness.

### 1.2.2 Stress-Flow Characteristics of Fracture Rock Mass

In recent years, many studies have been conducted to investigate the effects of stress path, seepage pressure, and fracture roughness on the permeability of fractured rock mass (Vogler et al., 2016; Wang et al., 2016a; Yang et al., 2018a). In laboratory experiments, the stress seepage test usually employs a rock mass with a single rock fracture (SRF), which is the simplest seepage test method and also a benchmark for complex fracture systems. Stress, seepage pressure, and fracture roughness are the main factors affecting SRF permeability. Fracture surface roughness can be quantized by three-dimensional laser scanning (Singh et al., 2015), as shown in Fig. 1.7.

Coal mining has a significant effect on the stress environment. Owing to spatiotemporal variations of mining-induced stress, irregular fractures develop in the surrounding rock mass. Any change in fracture aperture caused by compaction and shear deformation significantly affects fracture permeability. Thus, there are many laboratory studies on the effects of stress on the permeability of SRF. Rock permeability with SRF reduces exponentially with increasing effective stress (external

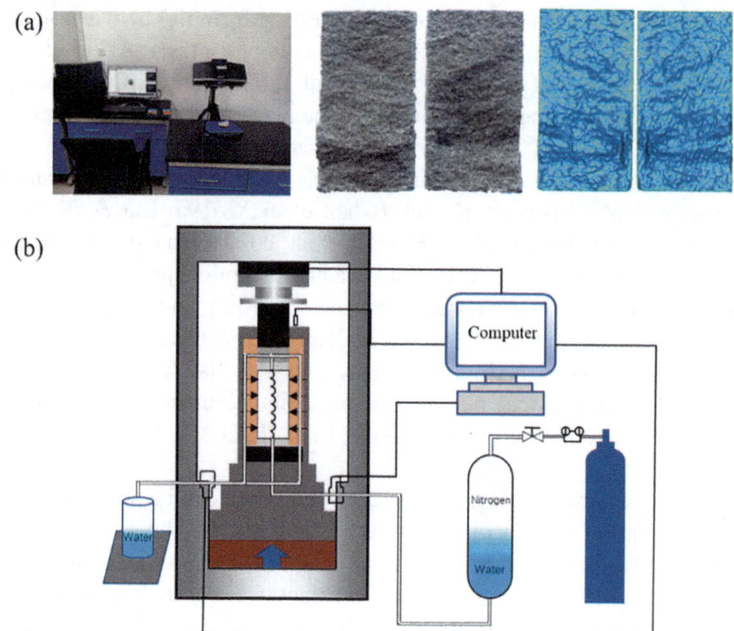


Fig. 1.7 3D scanning of fracture surface (**a**) and stress seepage test of SRF (**b**)

stress minus flow pressure) (Zhao et al., 2020c; Huo & Benson, 2016; Zhang et al., 2022c). Based on experimental tests, Meng et al. (2011, 2016) investigated the relationship between stress-strain and permeability for different rocks. They established a quantitative model to describe the relation between mining stress and permeability and also studied the dynamic variation law of rock stress and permeability coefficient. When loading a fractured rock mass, the asperities or particles within the fractures may undergo deformation, fragmentation, and re-arrangement, resulting in a sharp decrease in permeability. However, only the permeability loss caused by the elastic deformation of the asperities can be recovered during the unloading process (Zhang et al., 2018a), as shown in Fig. 1.8.

Fracture roughness is another crucial factor affecting SRF seepage. Fracture roughness depends on the rock properties (mineralogy), the fluid condition, the stress condition, and the fracture type (Kubeyev et al., 2022; Wang et al., 2023a). To study the influence of fracture roughness on seepage, it is necessary to characterize surface roughness. There are over 18 parameters proposed to describe the roughness, including four categories: fractal, random field, statistical, and empirical (Tse & Cruden, 1979; Wang et al., 2016b; Zhou et al., 2015). Among these, the joint roughness coefficient (JRC) is a widely used empirical quantity in geomechanics to characterize fracture roughness (Bhushan, 2000; Kim et al., 2013). At least four different ways can be used to estimate JRS, mainly based on the Barton and Choubey method and the Maerz method (Barton & Choubey, 1977; Grasselli & Egger, 2003; Maerz et al., 1990; Milne et al., 2009).

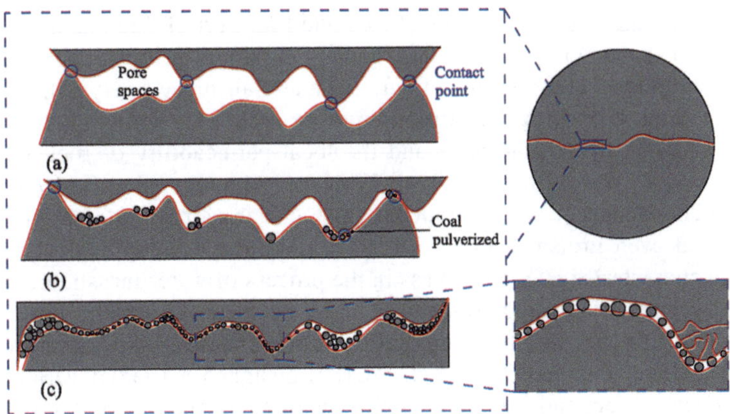


**Fig. 1.8** Simplified model of single persistent fracture compaction

Experimental evidence suggests that permeability assessment without considering the fracture roughness (i.e., parallel plates approximation) tends to overestimate permeability (Wang et al., 2016a; Zou et al., 2017). Huo et al. (2014) used CT scanning method to obtain fracture apertures under different stress states and calculated the permeability according to the parallel plates approximation. The calculated permeability was 2.6–4.9 times greater than the experimental values. Based on the lattice Boltzmann method and considering the factor of fracture surface roughness, Wang et al. (2017) put forward the modified formula of cubic law seepage on rough fracture surfaces. Test results by Ni et al. (2018) indicated that the seepage characteristics of a single fractured rock were greatly affected by the geometry of the fracture. The semi-empirical Forchheimer equation is established according to the variation of fracture pore size and roughness. In general, when the effective stress acting on the fractured rock is constant, the fracture permeability decreases with the increase in roughness (Babadagli et al., 2015; Kubeyev et al., 2022).

Seepage pressure, as another factor affecting single fracture seepage, reduces the effective stress acting on the fracture surface. In addition, the fluid pressure may also affect the fracture geometry. In particular, fracture expansion and inertia effects might occur under high-stress gradients (Zhou et al., 2016). The seepage pressure changes the strength and deformation resistance of the host rock matrix, resulting in changes in permeability with stress (Ma et al., 2013; Xiao et al., 2020).

### *1.2.3 Compaction-Flow Characteristics of Broken Rock Mass*

Based on permeability tests of cohesionless soil, Moffat and Fannin (2011) defined the characteristics of seepage instability, and proposed a method to calculate rock porosity and permeability under erosion. Richards and Reddy (2007)

believed that there are two stages of loss and transport of fine materials in rock mass under the water seepage effect, namely erosion and inrush. Erosion refers to that, under the water pressure effect, fine materials move freely within specific ranges without disturbing the surrounding rock particles, resulting in the redistribution or loss of fine materials, and the local permeability of rock mass will change slowly (Ranjith et al., 2014). With the continuous development of erosion, small substances are continuously lost, causing a rapid increase in permeability and even inrush disasters. To investigate the development of pore and seepage characteristics of rock mass in the process of water inrush, the relationship between porosity and permeability under water erosion conditions is studied by Ma et al. (2019), and the temporal evolution characteristics of permeability parameters are obtained. Liu et al. (2018a) designed a triaxial variable mass seepage test system and studied the influence of water pressure and initial porosity on erosion characteristics. They observed the transition of water flow state from Darcy to non-Darcy flow, and put forward the critical condition of water inrush. With the enrichment of monitoring means, CT and other perspective equipment are used to monitor the evolution of pore structure in the seepage process of broken rock mass. CT images indicated that fine particle erosion in the broken sample is obviously not homogeneous, which leads to the formation of seepage channels and uneven distribution of pore throats (Li et al., 2020d; Nguyen et al., 2019).

These studies mainly focus on the laboratory analysis of variable mass seepage of sandy soil. Regarding the water inrush problem of collapse column or fault under the action of complex stress fields in coal mining, Li et al. (2009), Wang et al. (2010), and Li et al. (2014) studied the whole process of water inrush in laboratory tests and revealed the water inrush mechanism and formation laws of water inrush channels. A time-varying law of water inrush from collapse columns under the mining influence was obtained. However, the structural characteristics of the collapse column itself have not been considered. On this basis, Wang et al. (2014), Kong and Wang (2018), Yu et al. (2018), and Zhang et al. (2019b) investigated variation laws of permeability and porosity with flow pressure considering the grading composition of collapse column filler, external water pressure, filling material, and rock block cementation. These works also uncovered the influence of mass loss rate and consolidated material on seepage characteristics in dynamic seepage system. Feng et al. (2018b) and Wu et al. (2019) considered the effect of Talbot power exponent $n$ on permeability, particle mass loss, and porosity, and discussed the permeability and water inrush parameters of broken rock mass under various Talbot grading. The relationship between broken rock mass loss and permeability was obtained.

Therefore, laboratory tests of broken rocks in a coal mining environment mainly focus on the effect of stress path, particle size, lithologies, and water flow rates on the variable mass seepage. AE monitoring, CT scan, and other means were used to observe changes in internal pore structures in the flow test, as shown in Fig. 1.9.

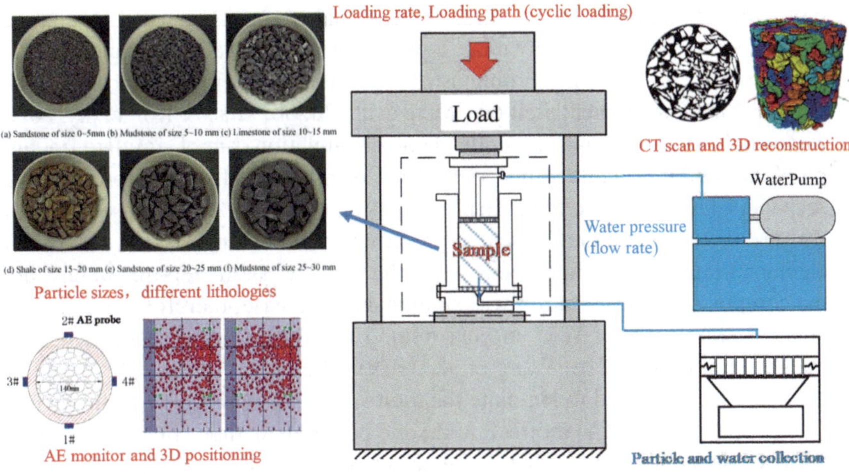


**Fig. 1.9** Seepage test of broken rock mass and related monitoring means

## 1.3 Simulation of WRI in a Coal Mining Environment

Numerical simulation is a promising method for WRI analysis in coal mining environments, especially for engineering-scale problems. According to different research purposes of WRI in coal mining, simulations can be divided into continuous rock mass simulation and broken rock mass simulation. Continuous rock mass simulation mainly includes roadway and coal pillar instability, mining water inrush, and other problems. Broken rock mass simulation includes water inrush from fault and collapse column, water storage, and drainage in the goaf caving zone.

### *1.3.1 Continuous Rock Mass Simulation*

The continuous rock mass simulation of WRI can be divided into five steps: choosing simulation methods, establishing an engineering-scale model, selecting constitutive models (water weakening should be considered), calibrating parameters (field-scale or laboratory-scale calibrations), and subsequent predictions and preventions.

1. Choosing simulation methods

   The analysis of WRI in coal mining belongs to engineering-scale simulation, which needs to consider the calculation efficiency. Finite element and discrete element simulation methods are frequently used. The finite element method has a wide range of applications, e.g., underground fluid, electromagnetic, heat transfer, and mechanics. (Bathe, 2007; Belytschko et al., 2009). Because the finite element method does not consider rock fractures, it makes the calculation more efficient and suitable for three-dimensional engineering-scale simulations (Chaskalovic, 2008; Erhunmwun & Ikponmwosa, 2017; Sabat & Kundu, 2021). Therefore, the engineering problems of WRI in coal mining are mostly calculated by the finite element method (Wang et al., 2020b; Zhao & Konietzky, 2020; Zhou et al., 2018b). However, finite element simulation cannot simulate fracture problems, and it is challenging to simulate the influence of fracture problems on seepage. Thus, the discrete element simulation method is usually used to solve fracture development and rock fracture problems involving WRI. The discrete element simulation method needs to set the fracture parameters separately, and the computation consumption will significantly increase; therefore, the discrete element simulation of WRI in coal mining mostly employs two-dimensional models (Bai et al., 2016; Wang et al., 2019). The indirect fluid-solid coupling method is mainly used to simulate the multi-physical field problems in the process of coal mining. The indirect fluid-solid coupling method solves each physical field separately and realizes mutual coupling by exchanging information among the fields. Compared with the fully coupled method, this approach could greatly improve computational efficiency.

2. Establishing an engineering-scale model

   Two problems need to be considered in the modeling process. The first is how to choose suitable element sizes of coal measure strata. Because of the large-scale models, if the element size is too fine, the model would be huge, and the computational efficiency may not be acceptable. If the element size is too large, the calculation accuracy might not be receivable. The second is to deal with the model boundary conditions. The boundary conditions should match the actual conditions as much as possible and enable the simulation to eliminate the boundary effect. Common WRI models in coal mining mainly include the roadway model, coal pillar model, and longwall face model, and some geological structures may be involved. In the process of roadway stability analysis under WRI, if the influence of longwall mining is not considered, the model is relatively small, and the model size should be as large as possible to avoid boundary effects (usually larger than five times the roadway size), as shown in Fig. 1.10a. In the coal pillar analysis, if the effects of mining are not considered, the coal pillar model size is also relatively small (Zhang et al., 2022a). Therefore, the aforementioned small models can employ uniform fine grids in the model. However, if the influence of longwall mining needs to be considered, the model should be carefully built to mimic the effects of longwall mining, as shown in Fig. 1.10b. In this case, only the zone of interest is subject to appropriate grid densification. Thus, the calculation accuracy could be guaranteed, and the computational consumption is also acceptable at the same time.

3. Selecting constitutive models

   Fluid-solid coupling analysis is needed in the analysis of WRI in coal mine simulations, including mechanical and seepage calculations. Mohr-Coulomb and strain softening failure criteria are commonly employed for mechanical calculation. Darcy's law is generally used for seepage calculation. In the coal

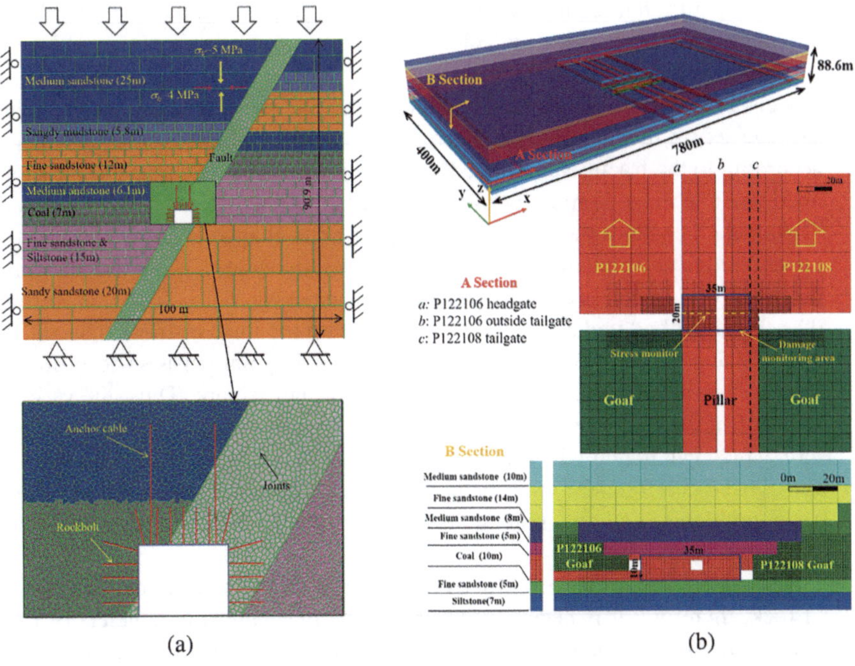


**Fig. 1.10** Typical models of WRI in a coal mine: (**a**) a 2D DEM roadway model involving a fault, (**b**) a 3D FEM coal pillar model considering the influence of longwall mining

mining WRI fluid-solid coupling calculation, two aspects should be considered (i.e., permeability updating and rock mass softening) due to the intense influence of mining and water softening.

Stress and damage/fracture are considered in permeability updating. Different permeability models are frequently employed to evaluate permeability changes in various rock mass (i.e., caving zone, fracture zone, and bedding zone) during longwall mining (Zhang et al., 2016c, 2020b). The permeability of rock mass in the bending subsidence zone can be updated according to the stress regime. In the fracture zone, the permeability varies based on the single fracture model, or the permeability directly changes according to field measurements or laboratory tests. The permeability of the broken rock in caving zone is relatively large, and it can be updated directly according to the strain, or according to the permeability model of broken rock mass (Zhang et al., 2018b, 2019c; Whittles et al., 2006; Rutqvist, 2011; Esterhuizen & Karacan, 2005, 2007). However, permeability models are usually derived from laboratory tests on specific coal and rock samples, which are generally unsuitable for all cases. In addition, the influences of stress loading and unloading on permeability changes are not considered in most permeability models (Zhang et al., 2020c); loading and unloading processes frequently occur in the field;

therefore, sufficient cautions are needed when employing these permeability models to simulate field-scale WRI problems.

The permeability methods mentioned above are mainly applicable to FEM models. For the discrete element simulation method, it is generally assumed that seepage exists in the contacts. Therefore, the permeability can be calculated according to the fracture aperture. There are two methods to update the fracture aperture. The most commonly used method is the fixed joint stiffness method. This method changes the aperture linearly with the change in stress and sets the upper and lower limits of the aperture. However, many tests show that the compressibility of fractures changes gradually (Chen et al., 2015, 2016). Therefore, the second method updates the joint aperture with the joint stiffness (Qiao et al., 2020; Zhang et al., 2022b). This method updates the joint stiffness according to the stress to match the actual compressibility of the fracture. Thus, the variable joint stiffness simulation method does not need to set the residual fracture aperture, which can match wider stress changes, but the computational efficiency is very low. Therefore, the first method is frequently employed in engineering-scale problems, and the second method is mainly used in laboratory-scale situations.

Laboratory research shows that WRI will reduce the mechanical parameters of rock; therefore, it is necessary to update the mechanical parameters of rock mass in the WRI simulation. Through the direct shear test and uniaxial compression test of rock mass with different water contents (or saturation level), the relationship between the weakening coefficient of rock mass and water content can be obtained. Therefore, as long as the moisture content of rock mass is obtained in the simulation, the mechanical parameters of the rock mass can be updated accordingly. There are three methods for calculating water content. (1) The first method is to update the water content through the updated permeability. Firstly, the porosity is obtained according to the cubic relationship between permeability and porosity. When the water pressure is greater than 0, it can be considered that the rock mass is saturated, and thus the water content can be calculated according to the porosity (Han et al., 2022a), as shown in Fig. 1.11. This method is suitable for saturated seepage. (2) The second method updates the rock mass strength according to the rock mass saturation (Pan et al., 2020), and this method is suitable for unsaturated seepage. (3) The third method is to calculate the water content according to the joint aperture and assume that water only exists in the joint (Wang et al., 2019). The first two methods usually apply to FEM models, and the third one usually applies to DEM models. The latter two methods are only applicable to the calculation of water content or saturation before failure. After failure, the rock mass strength is set as a residual value and will not be updated. In addition, the water content can also be updated according to the relation curve between the water content and the immersion time (Bai et al., 2016); however, the numerical time is difficult to correspond to the actual time.

The parameter calibration in the numerical simulation of WRI is basically consistent with other problems, including field-scale and laboratory-scale calibration.

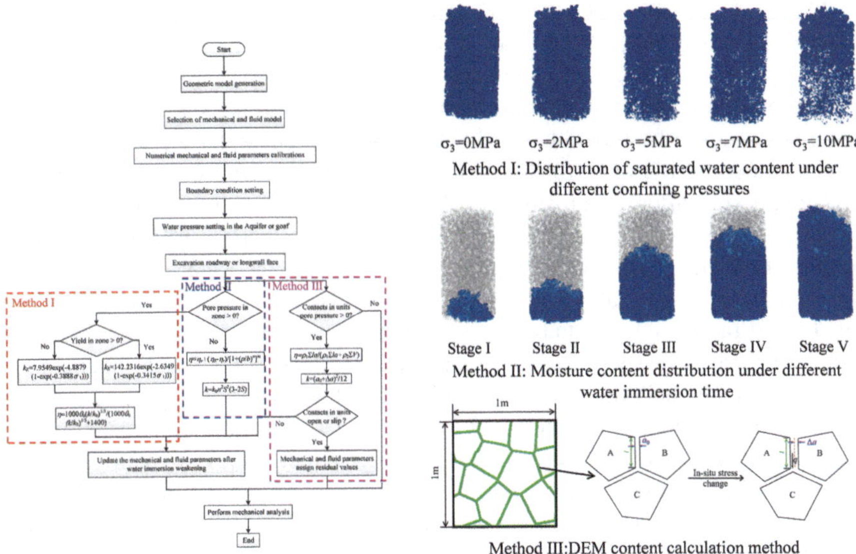


**Fig. 1.11** Three fluid-solid coupling methods with water softening and permeability updating. $\sigma_1$—effective stress; $\eta$—moisture content; $\eta_r$—residual volumetric water content; $\eta_S$—saturated volumetric water content; $k_S$—plastic zone permeability; $k_E$—elastic zone permeability; $\theta_0$—porosity before deformation; $k_0$—permeability before deformation; $k$—permeability after deformation; $S$—saturation; $p$—fluid pore pressure; $\rho_1$—fluid density; $\Sigma l$—sum of the contact lengths in the unit area; $\rho_2$—density of blocks in the unit area; $\Sigma V$—volume of blocks in the unit area; $a_0$—contact aperture at zero normal stress; and $\Delta a$—the contact normal displacement; $b$, $m$, and $n$—fitting parameters

After parameter calibration, subsequent predictions and preventions for the problems listed in Sect. 1.1 could be simulated. For example, the effect of mining on the seepage field was simulated by Zhang et al. (2019d) with FLAC3D–CFD (fluent) simulator to study the influence of mining on roof aquifer. Similarly, the mining-induced fluid influx evolutions, permeability variations, and strata stress changes are analyzed by Chen et al. (2019d) using COMSOL software. Zhang (2021) performed a coupling relationship among the floor failure, permeability evolution, and aquifer water uplift caused by mining using FLAC3D. Chi et al. (2019) used UDEC to simulate mining-induced uplift. Chen et al. (2022a) embedded a discrete fracture model in the process of simulating aquifer seepage to study the influence of fracture networks on roof water seepage. These simulations mainly consider the influence of mining stress and damage on water permeability, and then solve the problem of water inrush from the roof and floor. In addition, there are many water inrush simulations based on faults and collapse columns. However, in the simulations, faults and collapse columns are simplified as a continuum with specific strength and permeability rather than discontinuous broken coal and rock mass (Li & Xu, 2016).

The softening effect of water is usually neglected or simplified in the aforementioned analysis of WRI. Poulsen et al. (2014) considered the weakening effect of coal and rock in a saturated state on the coal pillar simulation by FLAC3D models. A similar simulation method considering the weakening of rock mass from dry to saturated is given by Fu et al. (2022). A time-dependent water weakening process was assumed by Bai et al. (2016) and implemented in a UDEC model to simulate the progressive failure of a roadway with water-rich roofs. However, the water seepage was not considered in these studies. Pan et al. (2020) considered the water weakening using the method II in Fig. 1.11 to simulate the water-weakening effect of rainfall on slope stability by FLAC3D models. Wang et al. (2019) used the method III in Fig. 1.11 to study the weakening effects on coal pillar stability in the goaf underground reservoir by UDEC models. However, two-dimensional simulation models are challenging to fully realize the influence of mining processes. Han et al., (2022a) adopted the method I in Fig. 1.11 to update the permeability and mechanical parameters for coal pillar damage characteristics affected by mining and water immersion by FLAC3D models. Besides, ground surface uplift can be observed when the mine goaf is flooded by groundwater as a result of water pressure changes (Dudek et al., 2020; Vervoort, 2021). For this problem, numerical simulation of flooding-induced uplift for abandoned coal mines was carried out by Zhao et al. (2021a), and the effect of water pressure caused by groundwater level recovery on the surface uplift is simulated by FLAC3D models. However, the WRI in this simulation is the supporting effect of water pressure on the rock stratum, not involving strength weakening and permeability updating.

## 1.3.2 Broken Rock Mass Simulation

1. Water weakening simulation

   Because few attempts have been conducted to test softening behavior of broken rock mass, it is difficult to simulate the softening effect of broken particles with WRI. On the one hand, broken rock mass cannot be further broken in the engineering-scale simulation, which eliminates the need to consider the water weakening on particle strength. On the other hand, the mechanical parameters of broken rock mass are greatly affected by the particle size and shape (Huillca et al., 2021; Zhang et al., 2020b, 2022e, 2022f). The softening coefficients of rock mass vary with particle sizes and shapes, so it is challenging to simulate their softening behavior in the numerical model. In addition, the fluid-solid coupling method simulates fluid elements and particle elements separately; that means there is no water content in the particles by default, so the water content cannot be calculated.

   Therefore, only the softening of particle contact parameters can be considered in the simulation of WRI in the broken rock mass. During the wetting process of broken particles, the friction coefficient between particles and the normal and shear stiffness of the contact gradually weaken. Baud et al. (2000) suggested that the friction coefficient was assumed to decrease by 10%. Other parameters can be obtained by matching mechanical experiments. It was found that the influence of water on the deformation of large particle assemblies is greater than that of small particle assemblies, especially under an environment with a high-stress ratio (ratio between major and minor principal stresses) (Shao et al., 2020).

2. Fluid-solid coupling simulation for broken rock mass

   CFD-DEM coupling method can well simulate granular systems distributed in fluid (Chen et al., 2021b; Shimizu, 2011; Bai et al., 2019b; Jiang et al., 2014; Climent et al., 2014). In CFD-DEM simulation, Newton's second law is used to describe the particle movement in the fluid; the fluid force is assumed to act at the center of the particle, and change its position and velocity (Cundall & Strack, 1979). Viscous damping is set to dissipate energy caused by particle collision (Jiang et al., 2005). The fluid flow is controlled by the Navier-Stokes equation (Kloss et al., 2012; Zhou et al., 2010). Virtual mass force, uplift forces (buoyancy), surface stress (differential fluid pressure and viscous force), and drag force are the main interaction forces between particles and fluid. The pressure gradient equation used in the simulation is determined by Reynolds number: Darcy equation with Reynolds number less than 10 and Ergun equation with Reynolds number > 10 (Bai et al., 2021; Guyer et al., 2009). The calculation method of drag force between particles and fluid is given by Di Felice (1994) and Xu and Yu (1997). Particle discrete element software, such as PFC, can be used for DEM parts simulation. Python scripting can be used for CFD parts and DEM-CFD interaction module simulation, as shown in Fig. 1.12. The DEM is used for the mechanical calculation of particles and provides interaction forces, porosity, particle velocity, and particle parameters to the fluid system. The CFD solver is used for flow calculation and provides fluid parameters to the DEM.

   However, in CFD-DEM simulation, the fluid element size is generally more than five times the particle size (Zhang et al., 2021a), which cannot accurately describe the fluid flow at the pore scale. Besides, the fluid element mesh cannot be changed during the simulation process, making it impossible to simulate the influence of external stress (Cui et al., 2016; Kruggel-Emden et al., 2016).

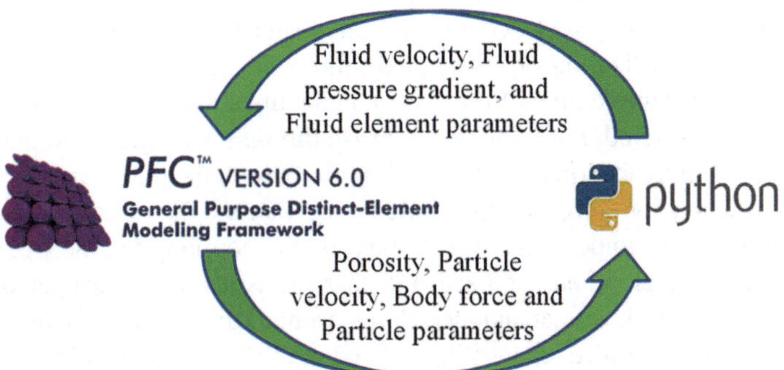


**Fig. 1.12** CFD-DEM coupling method by PFC and the Python solver

Lattice Boltzmann Method (LBM) is another DEM to simulate fluid-particle interactions, and it can be used for pore-scale fluid flow simulation, such as sand production and proppant compaction (Al-Shaaibi et al., 2013; Brumby et al., 2015; Cui et al., 2014; Fan et al., 2018; Han & Cundall, 2011). However, this method is computationally intensive and can only simulate 3D models with a few thousand particles (Zhang et al., 2020b). Therefore, this method is not suitable for engineering-scale simulations. Beyond CFD and LBM, other methods, such as Representative Elementary Volumes (REV) , Dynamic Fluid Mesh (DFM) method (Zhang et al., 2020b), Smoothed Particle Hydrodynamics (SPH) (Holmes et al., 2011), and Pore Network Method (Zhang et al., 2017a, 2019e) have been proposed to solve various fluid-particle interaction issues with DEM.

However, there are few fluid-solid coupling methods for simulating WRI of broken rock mass in a coal mining environment. It is only used in the simulation of pore evolution of broken rock mass in caving zone by water storage and drainage of underground reservoirs (Zhang et al., 2021f) with a laboratory-scale dimension because broken rock mass simulations in coal mining usually have a vast scale (hundreds of meters) with a very large size distribution scope (from microns to dozens of meters), the fluid-solid coupling simulation of thus a complicated system usually needs huge calculation consumption. In future work, these fluid-solid coupling simulation methods (e.g., CFD-DEM and LBM–DEM) of broken particles should be used for reference to study the water inrush of fault and collapse column, and the fluid-solid coupling in the goaf reservoir.

## 1.4 Concluding Remarks and Prospects

Mine water disaster is regarded as one of the most critical problems in coal mining and has always been a research hotspot in coal mining. Meanwhile, mine water is also a resource, and some mining areas with water shortages focus on the study of rational utilization and protection of mine water. In coal mining, the research related to WRI mainly focuses on the water inrush from the roof and floor, water weakening of roadway surrounding rock (coal pillar), the water inrush from fault or collapse column, and the utilization of abandoned mines. For the stability evaluation, the influence of water weakening of rock mass on the instability of roadway (including coal pillar) was studied. For mine water inrush through geological anomalies (fault or collapse column), existing research mainly focuses on the judgment conditions of water inrush and its corresponding influencing factors. For water seepage from mine roof and floor under mining disturbances, the coupling law between mining stress, fracture development, and seepage characteristics was investigated. For the underground reservoir and abandoned flooded mine, the research mainly involves the influence of water level change in goaf and the WRI on surrounding rock stability, surface deformation, and pore change in the caving zone. To solve these problems, laboratory tests and numerical simulations are the two most commonly used means. Laboratory tests provide mechanical and seepage parameters for numerical simulations; permeability updating and mechanical weakening in the process of numerical simulation need to be determined by laboratory tests. Numerical simulations can apply the laboratory results in coal mining and realize the scale-up application of laboratory-scale results.

Laboratory tests mainly involve three types of WRI: intact rock mass, fractured rock mass, and broken rock mass. The WRI test of broken rock mass is mainly used to study the seepage of water inrush from faults and collapse columns, focusing on the variable mass seepage within the broken rock mass. These studies consider the effects of the particle size and lithologies, water pressure, water velocity, and external stress environment on the porosity/permeability of broken rock mass, so as to obtain the threshold of water inrush from faults or collapse columns. In addition, the changes in porosity and permeability of broken rock mass in the caving zone also involve variable mass seepage in the process of water storage and drainage of underground reservoirs. AE monitoring, CT scan, and other means were used to observe the changes in internal pore structures in the flow test. It is concluded that variable mass flow is the main reason for the changes in porosity and permeability of broken rock mass. External loading limits variable mass flow, while internal water pressure promotes variable mass flow. The lithology of broken rock affects the water weakening degree. The rock mass with small particle size is the main reason for the change in porosity in the process of seepage.

The WRI test of fractured rock mass is mainly used to study the water inrush from fractured rock mass in the roof and floor. The strong mining stress disturbance of the coal seam causes the fracture zone in the roof and floor, and water inrush disasters may occur when the fracture develops into aquifers. Thus, the research mainly focuses on the coupling characteristics of stress-fracture-seepage of the fractured rock mass. In addition to stress and flow pressure, the roughness of the fracture surface is also an interesting research field. Fracture surface morphology can be obtained by 3D laser scanning, and its roughness can be characterized by JRC. Loading makes the permeability of fractured rock decrease exponentially and destroys the original fracture contact surface; therefore, the permeability cannot be fully recovered during unloading. Under the same stress condition, the greater the roughness of the fracture surface, the smaller the permeability. The seepage pressure changes the strength and deformation resistance of a rock surface, resulting in a change in permeability with stress loading.

The WRI test of intact rock mass is mainly concentrated on water weakening and its effect on the instability of the roadway and coal pillar. The research on the water weakening of rock mass particularly emphasizes the effects of water content, water immersion time, and cyclic water immersion. The strength degradation of rock mass varies with water content, immersion time, and WD cycles, which are related to the type of rock mass and mineral composition. Generally, the strength decreases exponentially with the increase in water content. The first several WD cycles have a significant impact on the strength degradation, and the strength degradation gradually weakens with the increase in WD cycles. With the increase in immersion time, the degree of rock weakening decreases gradually. The corresponding AE count and intensity and the infrared radiation intensity also weaken accordingly. WRI enhances the plasticity of rocks and reduces brittleness.

Most common experiments usually use a loading regime (monotonically increasing the load until failure occurs) to investigate rock performance and to test mechanical and hydraulic parameters. However, some studies have shown that rock behaviors depend on the applied stress path (e.g., Zhang et al., 2019a, 2020a). Therefore, it is still doubtful whether the tested parameters could accurately reflect field rock behaviors because laboratory tests employ a stress path that significantly differs from that infield practice. Several studies have shown that employing the field stress path in lab experiments could more accurately reproduce the field rock behaviors (Bai et al., 2019b; Zhang et al., 2016e, 2020a). Therefore, it is recommended to consider the stress path effect when studying the WRI. The stress path can be obtained from field monitoring or numerical simulations.

The numerical simulation analysis of WRI in coal mining can be divided into two categories. One is the simulation analysis of continuous rock mass, including the softening analysis of roadway and coal pillar when encountering water, and the analysis of water inrush from coal seam roof and floor caused by mining fractures. The other is the simulation analysis of broken rock mass, which is mainly applicable to the analysis of water storage and drainage in the goaf and the water inrush from faults and collapse columns. However, water inrush from faults and collapse columns is still mainly based on continuous rock mass simulations. According to different research problems, the size and mesh division of the models of coal mine WRI also vary. In general, the model should consider the effect of longwall mining, indicating that the model would be very large (hundreds of meters). Permeability variation and rock softening are two main problems to be considered in continuous rock mass simulation. The permeability can be updated based on stress (or strain) and fracture (or plastic zone). Rock softening can be updated according to rock saturation, water content, or calculation time. Saturation can be calculated according to water pressure, and water content can be updated according to porosity (permeability) or fracture aperture. Finally, coal and rock mass parameters can be updated through the relationship between water content (saturation) and rock mass strength. In the simulation of WRI, the indirect fluid-solid coupling method is generally used. The permeability is updated after the mechanical calculation, and the rock mass strength is updated after the seepage calculation.

Few attempts focus on the WRI simulation of broken rock mass and the associated strength degradation in a coal mining environment while most studies are inclined to the interaction between water seepage and broken rock mass. At present, the CFD-DEM method is mainly used in the fluid-solid coupling analysis of broken rock mass. The particle element and fluid element are calculated separately, and the coupling analysis is carried out through the interaction between particles and water. The DEM is used for the mechanical calculation of particles and provides interaction forces, porosity, particle velocity, and particle parameters to the fluid system. The CFD solver is used for flow calculation and provides fluid parameters to the

DEM. However, in the process of CFD simulation, the fluid element mesh cannot be deformed, and the influence of external stress on the fluid mesh cannot be simulated. Because the fluid element size is larger than the particle size, the CFD-DEM simulation method cannot describe the pore-scale fluid flow. LBM is another method to simulate fluid-particle interactions; it can be used for pore-scale fluid flow simulation. However, this method needs a large amount of calculation consumption and is challenging for engineering-scale analysis. These fluid-solid coupling methods of broken rock are mainly used for the analysis of WRI in sand and soil and are rarely used in coal mining problems.

Nowadays, in addition to the traditional water disaster prevention and water immersion weakening of rock mass, the research in coal mining turns to the problem of WRI in the reuse of abandoned mines. Especially the issue of WRI in the construction of mine underground reservoir for pumped storage power stations. Many experimental methods and advanced observational and monitoring means have been proposed to study the variation of rock mass behavior and the associated micromechanisms during the WRI process in various coal mining environments. The experimental research mainly adopts specific rock mass (intact, fracture, and broken) or water environment under various stress environments according to specific WRI problems in coal mining and uses a variety of micro-characterization means to interpret the water-rock weakening mechanism. However, more attention should be paid to how the laboratory-scale results are applied to engineering-scale problems. Therefore, some qualitative conclusions need to be transformed into quantitative models, which can be applied to field-scale models. In addition, the factor of immersion time is limited to a short time in laboratory tests (usually less than half a year); however, the mechanical parameters of rock are weakened under long-term water immersion (more than several years), and the evolution of pore structure should pay more attention. Besides, for broken rock mass, the influence of water immersion on its strength degradation can be further studied. For single fracture rock mass, the influence of water immersion or seepage on the roughness of fracture highlights a critical need for further fundamental studies.

At present, the numerical simulation can well mimic the weakening and instability of roadway and coal pillar by water immersion and the water inrush from coal seam roof and floor. The effects of stress, fracture development, water pressure, and saturation (water content) on the strength and permeability of coal and rock mass can be considered in the simulation process. However, considering the calculation efficiency, the simulation at this stage is basically an indirect fluid-solid coupling method. The complete fluid-solid coupling methods can be tried in small and simple models. Besides, the influence of water immersion time on strength is not considered in the simulation of WRI. Moreover, there are few fluid-solid coupling analyses for broken rock mass. In further studies, the simulation methods such as CFD-DEM can be used to study water inrush problems in the fault or collapse column of coal mining.

BY NC ND

# Chapter 2
# Strength Weakening and Its Micromechanism in Water–Rock Interaction

## 2.1 Introduction

Water-rock interactions (WRIs) are a topic of interest in geology and geotechnical engineering. Many physical and chemical reactions are involved in the WRI, including lubrication, precipitation, oxidation-reduction, and ion exchange. Many geological hazards and engineering failures, such as slope stability (Zhao et al., 2018d), reservoir dam stability (Ukpai, 2021), rock bursting (Chen et al., 2019a), karst collapse (Bai et al., 2013), and water inrush from mines and tunnels (Huang et al., 2016; Li et al., 2019b), are caused by the WRI. Analysis and interpretation of the influence of water on the mechanical behavior of rocks are based on the above problems. In recent years, many studies have illustrated the effects of water on the mechanical characteristics of rocks. In general, the presence of water reduces the elastic modulus, compressive strength, cohesion, tensile strength, and rock brittleness (Baud et al., 2000; Erguler & Ulusay, 2009; Zhou et al., 2017; Talesnick & Shehadeh, 2007) and changes the fracture distribution and fragment shape after rock failure (Haberfield & Johnston, 1990; Shen et al., 2020; Guha Roy et al., 2017; Kataoka et al., 2015). The water presence increases the pore water pressure in rock and soil mass, reducing the effective stress of rock and soil mass skeleton particles and changing the physical and mechanical parameters of the rock and soil mass. In addition, it is easy to cause the dissolution or change of mineral composition in rock, soil mass, and cement between grains to produce new mineral composition. The WRI effect will be greater if the water contains corrosive mineral components (Luo et al., 2021). For example, rocks with higher clay mineral content are more susceptible to water penetration (Verstrynge et al., 2014).

This short review focuses on the water weakening of rock strength and its influencing factors, such as the water content, immersion time, and wetting-drying cycles. On this basis, the micromechanism of WRI and its corresponding research methods are addressed.



C. Zhang, *Water Rock Interaction in Underground Coal Mining*,
https://doi.org/10.1007/978-981-95-9957-8_2

## 2.2 Water Weakening Characteristics and Its Influencing Factors

Mechanical experiments combined with corresponding characterization methods are the primary methods used to study the weakening characteristics of the WRI. In conventional mechanical experiments (uniaxial and triaxial compression tests, Brazilian splitting tests, shear tests, point load tests, needle penetration tests, and Hopkins impact tests, as shown in Fig. 2.1), the shear, tensile, compressive, hardness, and dynamic load strengths can be obtained. In the test process, in addition to obtaining the stress-strain curves, acoustic emission (AE), infrared radiation temperature, digital image correlation (DIC), 3D laser scanning, and other methods are typically used to characterize the influence of the WRI (Fig. 2.1).

The weakening analysis of the WRI focuses on the influence of water content, immersion time, and wetting-drying cycles on rock strength, as well as the corresponding AE and infrared radiation (IR) characteristics.

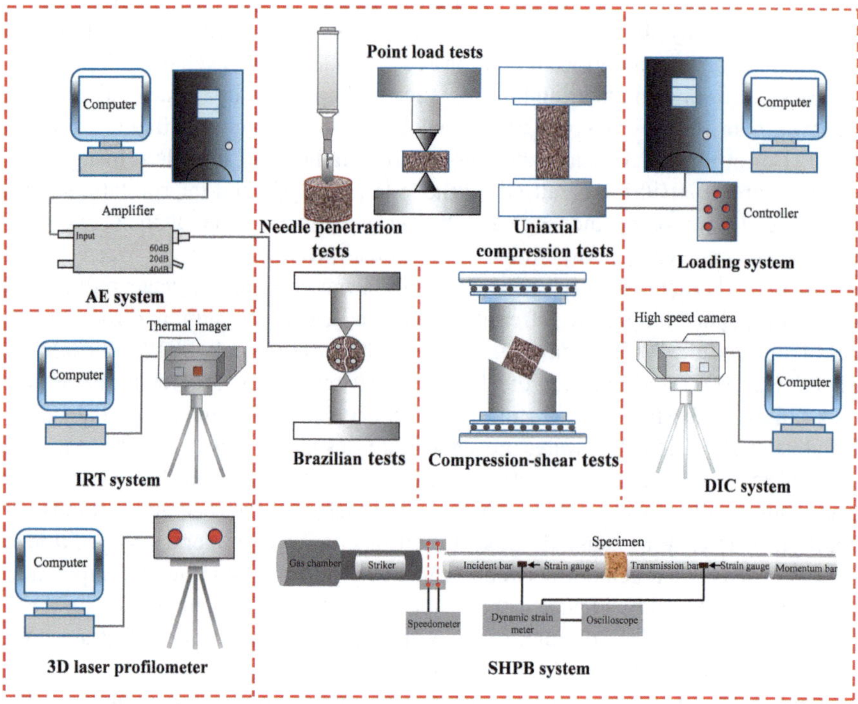


**Fig. 2.1** WRI mechanics test and primary analysis

### 2.2.1 Water Content

Water content can be divided into two aspects: (1) the water content state, including the dry state, natural water content state, and saturated water content state, and (2) the actual moisture content of the rock mass. There are two methods for preparing rock samples with different water contents: direct immersion in water and a non-destructive immersion method that places rock samples in a wet and closed environment to avoid the exchange or hydrochemical reaction between rock and water.

The compressive, shear, tensile strength, and elastic modulus decrease to varying degrees with the increasing water content, as summarized in Table 2.1. Among these, mudstone rocks have the highest water sensitivity. Granite and other hard rocks have the lowest sensitivity to water, and some rocks are unaffected by water. An increase in water content enhances the plasticity of the rock mass and reduces brittleness (Noël et al., 2021). Some quantitative relationships (fitting formula) between the mechanical parameters ($\sigma_t$, $\sigma_c$, $\sigma_{tc}$, $\tau$, $E$, $v$, $c$ $\varphi$, etc.) and water content are summarized in Table 2.1. In addition, the water content reduces the fracture toughness after rock failure (Zhou et al., 2018b, 2018d; Hua et al., 2015).

AE describes the energy release characteristics of rocks during fracturing or cracking. Hard rocks typically accumulate more energy during loading and would release stronger AE signals during failure. As the water content increases, the cumulative and peak AE signals, high-frequency AE signals, AE signal differences, and AE signal distribution uniformity decrease (Li et al., 2019a, 2021b; Ranjith et al., 2008; Guo et al., 2018; Lin et al., 2019; Zhu et al., 2020b; Liu et al., 2019). In addition, with an increase in water content, AE is concentrated in the fracture stage, and the fractal dimension decreases (Song et al., 2020; Kong et al., 2017, 2019). Wet rocks generally produce small fragments during failure, which decreases the fractal dimension, and irregular fractures occur during loading.

For the correlation between moisture content and IR, the average IR temperature (AIRT) generally increases with increasing load and depends on the state of the rock samples (wet or dry, damaged or not). Before the damage, wet rocks generally show a faster increase in the AIRT, whereas dry rocks would produce more increments in the AIRT during damage (Deng et al., 1997; Liu et al., 2010). However, in the entire loading process, the higher the water content, the greater the AIRT, and the smaller the AIRT fluctuation (Zhou et al., 2018e). Sun et al. (2021a) further showed that the applied stress controls IR; for example, the IR count (IRC) will simultaneously increase when the stress suddenly drops. The water content reduces the stress due to water weakening. The above conclusions mainly rely on the uniaxial compression tests. Shear tests by Yao et al. (2020a) concluded that the IR characteristics were comparable to those of the uniaxial compression tests.

With increasing water content, the dynamic load strength and dissipated energy decreased, but the elastic modulus increased. The dynamic strength of saturated rocks is more sensitive to the strain rate than that of dry rocks. With an increase in the strain rate (43.9–156.7 $s^{-1}$), the water weakening effect decreases gradually (Cai et al., 2020b). Further, compared with the uniaxial compressive strength of dry rock

**Table 2.1** Summary of laboratory tests of the effect of different water contents on the physical and mechanical characteristics of coal and rock

| Rock type (Region) | Water content | Experimental equipment | Mechanical parameters | Change trend with water content | Reference |
|---|---|---|---|---|---|
| Sandstone (Tennessee, USA) | $\omega$: Dry, saturated | TTA, SEM | $\sigma_{tc}$ | $\sigma_{tc}$ decrease | Feucht and Logan (1990) |
| Coal (Ningxia, China) | $\omega$: 0%, 7.10%, 15.68%, 22.90%, 23.09% | CS tests, AE system, IRM system | $c$, $\varphi$, $\tau$, amplitude of the AIRT | $\varphi = 2.39e^{-0.17\omega} + 37.24$, $c = 0.371e^{-0.12\omega} + 0.8898$, $\tau = -0.543\omega + 3.7797$, the amplitude of the AIRT decrease | Yao et al. (2020a) |
| Granite, tuff, andesite (Fukushima, Japan) | Dry, saturated | UTS tests | $\sigma_c$, $\sigma_t$, $R_{UTS}$, $R_{UCS}$ | $\sigma_c$ and $\sigma_t$ decrease, $R_{UTS} = 1.61R_{UCS}$ | Hashiba and Fukui (2015) |
| Sandstone (Sichuan, China) | $\omega$: 0%, 1.03%, 2.04%, 3.14% | UCS tests, AE system, SEM | $\sigma_c$, $E$ | $\sigma_c$ and $E$ decrease | Li et al. (2021) |
| Black sandstone (Sichuan, China) | $\omega$: 0%, 0.32%, 0.34%, 0.70%, 0.91%, 1.18%, 1.62% | UCS tests | $\sigma_c$, $E$, $v$ | $\sigma_c = 80.604*e^{-0.9044\omega} + 43.17$<br>$E = 20.451–4.7481\omega$<br>$v = 0.2 + 0.011e^{1.748\omega}$<br>$\varepsilon_c = 0.1213\omega^2–0.2744\omega + 0.7216$ | Tang et al. (2018) |
| Coal (Heilongjiang, China) | Dry(0%), natural(0.43%), saturated(1.70%) | MTTA, AE system, SEM, HSVC | $\sigma_{tc}$, $E$, cumulative AE counts, cumulative AE energy | $\sigma_{tc}$ and $E$ decrease, cumulative AE energy and counts decrease | Sun et al. (2016) |
| Coal (Henan, China) | $\omega$: 0%, 6.00%, 9.75%, 10.96% | UCS, AE system, MIP, XRD, SEM | $\sigma_c$, RA, AF | RA increase, $\sigma_c$ and AF decrease | Yao et al. (2019a, 2019b) |
| Shale beams (Pennsylvania, USA) | Dry(0%), saturated(%) | TPB, CT, AE system | $\sigma_t$ | $\sigma_t$ decrease | Lu et al. (2021) |
| Mudstone (Xinjiang, China) | $\omega$: 0%, 1.00%, 2.00%, 3.00%, 4.00%, 5.10%, 6.00% | UCS, UTS, SEM | $\sigma_c$, $\sigma_t$, $c$, $\sigma_{tc}$, $\varepsilon_c$ | $\sigma_c = 19.32e^{-0.52\omega} + 4.06$, $\sigma_t = 2.42e^{-0.44\omega} + 0.29$, $c = 5.22e^{-0.59\omega} + 1.63$, $\sigma_{tc}$ decrease, $\varepsilon_c$ decrease and then increase | Liu et al. (2021) |

| Rock type (Region) | Water content | Experimental equipment | Mechanical parameters | Change trend with water content | Reference |
|---|---|---|---|---|---|
| Sandstone (Shandong, China) | $\omega$: 0%, 0.8%, 1.6%, 2.4%, 3.2% | XRD, FJRE system | $\sigma_c$, $E$, $D_c$, | $\sigma_c = 52.5416e^{-1.9115\omega} + 37.6738$, $E = 5.9348e^{-1.5541\omega} + 7.5865$, $D_c = 0.6445–0.6445e^{-0.8253\omega}$ | Sun et al. (2021c) |
| Tuff (Sapporo, Japan), sandstone (Kushiro, Japan) | $\omega$: Dry, saturated | XRD, NP | $V_p$, NPI | $V_p$ and NPI decrease | Fujii et al. (2020) |
| Siltstone and gypsiferous rock (Alicante, Spain) | $\omega$: Dry, saturated | UCS, XRD, MIP | $\rho$, $\sigma_c$, NPI, $E$ | $\rho$ increase, $\sigma_c$, $E$ and NPI decrease, $\sigma_c = 0.13389$NPI (siltstone), $\sigma_c = 0.12559$NPI (gypsiferous rock) | Rabat et al. (2020) |
| Sandstone (Shanxi, China) | $\omega$: 0%, 1.13%, 1.58%, 2.52%, 3.63% | UCS, SEM, XRD, creep test | $\sigma_c$, $E$, $t_1$, $\varepsilon'$ | $\sigma_c = 44.600e^{-0.399\omega} + 66.6602$, $E = 5.640e^{-0.357\omega} + 13.426$, $t_1$ increase, $\varepsilon'$ increase | Chen et al. (2021b) |
| Clay-rich sandstone (Gansu, China) | $\omega$: 0.69%, 0.88%, 1.00%, 1.73%, 1.99%, 2.24%, 2.28% | XRD, SEM, MI test, SV test, SD test | $E$, $h$, $H$, $V_p$ | $h$ and $V_p$ increase, $E$ and $H$ decrease | Azhar et al. (2020) |
| Red sandstone (Hunan, China) | $\omega$: Dry, saturated (3.40%) | SEM, UCS | $E$, $\sigma_c$, $t_1$, $\varepsilon'$ | $\sigma_c = 55.21e^{-0.7502\omega} + 51.6$, $E = 6.183e^{-0.6847\omega} + 10.62$, $t_1$ increase, $\varepsilon'$ increase | Tang et al. (2018) |
| Coal (Shanxi, China) | $\omega$: 0%, 2.37%, 3.78%, 5.29% | UCS, AE system | $E$, $\sigma_c$ | $\sigma_c = -2.56\omega + 23.25$, $E = 1.05e^{-0.50\omega} + 0.72$ | Yao et al. (2016) |
| Coal–rock combination (Shanxi, China) | $\omega$: Dry, natural, saturated | AE system, IRM system | $E$,$\sigma_c$, $\varepsilon_c$ | $\sigma_c$, $E$, and $\varepsilon_c$ decrease | Yao et al. (2020b) |
| Coal (Sichuan, China) | $\omega$: Dry, saturated | MTTA, CT, AE system | $E$, $\sigma_{tc}$, $\varepsilon_c$, RA, AF | $\sigma_{tc}$, $E$, AF, and $\varepsilon_c$ decrease, RA increase | Liu et al. (2019) |
| Sandstone (China) | $\omega$: 0%, 0.50%, 1.50%, 2.50% | UCS, IRM system | $\sigma_c$, $\varepsilon_c$, AIRC | $\sigma_c$, $\varepsilon_c$ decrease, AICR increase | Sun et al. (2021b) |

(continued)

**Table 2.1** (continued)

| Rock type (Region) | Water content | Experimental equipment | Mechanical parameters | Change trend with water content | Reference |
|---|---|---|---|---|---|
| Red sandstone (China) | $\omega$: Natural (0.65%), saturated (2.37%) | UCS, MTTA, XRD | $E$, $V_p$, $\sigma_c$ | $V_p$ increase, $\sigma_c$, and $E$ decrease | Luo (2020) |
| Marble, limestone (Sichuan, China) | $\omega$: Dry (0%), saturated (marble: 3.01%, limestone:1.87%) | UCS, AE system, XRD | $E$, $\sigma_c$, H-type waveforms, L-type waveforms | $\sigma_c$ and $E$ decrease, H-type waveforms decrease, L-type waveforms increase | Zhu et al. (2021) |
| Red sandstone (Hunan, China) | $\omega$: 0%, 0.70%, 1.60%, 2.60%, 3.50%, 4.60%, 4.70% | UCS, SEM, XRD | $\sigma_c$, $E$, $\varepsilon'$, $t_1$ | $\sigma_c = 24.4e^{-0.55\omega} + 49.4$, $E = 3.4e^{-0.34\omega} + 7.4$, $\varepsilon'$ increase, $t_1$ decrease | Yu et al. (2019) |

*AE* acoustic emission, *CT* computed tomography, *NP* needle penetration, *CS* compression shear, *UTS* unconfined tensile strength, *UCS* unconfined compression strength, *MTTA* modified true triaxial apparatus, *FJRE* five joint rheological experimental, *TPB* three-point bending, *MI* micro-indentation, *SV* sonic velocity, *SD* slake durability, *SEM* scanning electron microscopy, $\varepsilon_c$ peak strain, *HSVC* high speed video camera, *IRM* infrared radiation monitoring, *MIP* mercury intrusion porosimetry, *RA* rise time divided by peak amplitude, *AF* counts divided by duration, $\varphi$ friction angle, $c$ cohesion, $E$ Young's modulus, $\sigma_t$ tension strength, *NPI* needle penetration index, $\sigma_c$ uniaxial compressive strength, $\sigma_{tc}$ triaxial compressive strength, $\tau$ shear strength, $v$ Poisson's ratio, $\rho$ density, $V_p$ P-wave velocity, $t_1$ creep time-to-failure, $\varepsilon'$ minimum creep strain rate, $h$ indenter depth, $H$ hardness, *AIRC* average infrared radiation count, $R_{UTS}$ reduction of uniaxial tension strength, $R_{UCS}$ reduction of uniaxial compression strength samples, that of saturated rock increases with the loading rate in two stages: rapidly increases at low loading rates, and then decreases at high loading rates (Zhu et al., 2021).

### 2.2.2 *Wetting-Drying Cycles*

Many practical engineering problems involve wetting-drying (WD) cycles, such as rocks in exposed slopes, coastlines, and pumping reservoirs. Similar to preparing rock masses with different water contents, there are also two methods (free and pressure immersion method, air and oven (or heater) drying) for preparing rock samples experiencing WD cycles. A vacuum pressure condition is adopted to accelerate rock saturation when preparing rock immersion, whereas the free immersion method is performed under atmospheric pressure conditions. In the drying process, oven drying can accelerate the drying of rocks compared with air drying. Previous laboratory tests have shown that the WD cycle treatment has a significant impact on the mechanical and physical properties ($A_w$, $P$, PLI, SDI, $K_{IC}$, $K_{IIC}$, $K_{eff}$, $V_p$, $\sigma_c$, $\sigma_t$, $E$, $c$ $\varphi$, $\tau$, $\sigma_{cd}$, $E_d$, $T_c$, $R$, and $H$) of the rock, as summarized in Table 2.2 (Liu & Zhang, 2020; Momeni et al., 2017; Zhou et al., 2018b; Chen et al., 2019c; He et al., 2020b).

Although the WD cycle treatment methods were different in these studies, consistent weakening characteristics were observed. With progressive WD cycles, the porosity and water absorption increased monotonically, whereas the other parameters in Table 2.2 generally decreased. The decrease in mechanical and physical parameters gradually diminished with the progression of WD cycles (Sun & Zhang, 2019; Khanlari & Abdilor, 2015; Huang et al., 2010; Fu et al., 2017; Zhao et al., 2018a; Gratchev et al., 2019; Yao et al., 2019b; Wu et al., 2020b; Cai et al., 2020a; Li et al., 2021d). For example, it was found that the first 10 WD cycles significantly impacted the rock's strength; there was no change (Zhao et al., 2021b; Guo et al., 2021) in the following cycles. The fracture toughness and crack propagation were also affected by the WD cycle. The fracture energy and fraction coefficient decrease with WD cyclic treatment (Zhao et al., 2017b, 2017c; Song et al., 2019; Ma et al., 2018). The rock's mineral composition has been found to be the main factor affecting the weakening of the WD cycle (Zhou et al., 2017; Tang et al., 2021a).

### 2.2.3 *Immersion Time*

The immersion time is more closely related to the actual field situation than the water content. Underground rock masses are occasionally immersed in water for months or years, and immersion time significantly affects rock strength (Bai et al., 2016). However, in most laboratory tests, the maximum immersion time is generally no longer than 1 year. The immersion time is currently limited because some

**Table 2.2** Summary of laboratory test results of rock mechanical and physical properties with WD cycle treatment

| Rock type | Sample size | Treatment methods | | Cycle number | Test results | Reference |
|---|---|---|---|---|---|---|
| | | Wetting | Drying | | | |
| Coal | CS (Φ50 × 100) | WW (24 h) | OD (95 °C/24 h) AD (20 °C/4 h) | 10 | $A_w$ increase; $\sigma_c$, $E$ decrease | Chen et al. (2019b) |
| Granite | CS (Φ50 × 100) | WW (24 h) | OD (105 °C/12 h) | 60 | $\sigma_c$, $E$ decrease | Chen et al. (2019c) |
| Sandstone | CS (Φ50 × 100) | WW (24 h) | OD (105 °C/12 h) | 20 | $\sigma_c$, $E$, $c$, $\Phi$ decrease | Liu et al. (2018b) |
| Granite | CS (Φ50 × 100) | WW (24 h) | OD (105 °C/12 h) | 20 | $\sigma_c$, $E$, $c$, $\Phi$ decrease | Qin et al. (2018) |
| Sandstone; mudstone | CS (Φ50 × 100) | WW (24 h) | OD (110 °C/12 h) | 40 | $A_w$ increase; $\sigma_c$, $E$ decrease | Huang et al. (2018) |
| Greywacke; basalt | CS (Φ50 × 50) | WW (24 h) | OD (100 °C/12 h) | 40 | PLI, SDI decrease | Gratchev et al. (2019) |
| Weak muddy intercalation | CS (Φ61.8 × 20) | WW (24 h) | OD (40 °C/45 h) | 6 | $\tau$, $c$ decrease | He et al. (2020a) |
| Sandstone | CS (length to diameter ratio of 2.5–3.0) | WW (24 h) | OD (110 °C/12 h) | 40 | $\sigma_c$, decrease | Khanlari and Abdilor (2015) |
| Sandstone | CS (Φ25 × 50) | WW (24 h) | OD (80 °C/2 h) | 50 | $T_c$, $\sigma_t$, $V_P$ decrease | Sun and Zhang (2019) |
| Ignimbrite | CS (Φ54 × 108) | WW (24 h) | OD (105 °C/12 h) | 50 | $A_w$ $P$ increase; $V_p$, $\sigma_c$, decrease | Özbek (2014) |
| Granitoid rocks | Rock pieces | WW (24 h) | OD (110 °C/24 h) | 40 | SID decrease | Momeni et al. (2017) |
| Sandstone | CCBD specimen | WW (48 h) | OD (105 °C/24 h) | 7 | $K_{eff}$, $\sigma_t$ decrease | |
| Sandstone | CSTBD specimen | WW (48 h) | OD (105 °C/24 h) | 7 | $K_{IIC}$, decrease | |
| Sandstone | CCNBD specimen | WW (48 h) | OD (105 °C/24 h) | 20 | $P$ increase; $K_{IC}$, $K_{IIC}$ decrease | Dehestani et al. (2020) |
| Sandstone | CCBD specimen | WW (48 h) | OD (105 °C/24 h) | 7 | $K_{IC}$, $\sigma_t$ decrease | Hua et al. (2015) |
| Mudstone | Rock pieces | WW (3 d) | AD (26 °C/24 h) | 11 | SID, decrease | Liu and Zhang (2020) |

| Rock type | Sample size | Treatment methods | | Cycle number | Test results | Reference |
|---|---|---|---|---|---|---|
| | | Wetting | Drying | | | |
| Sandstone | CS (Φ50 × 100) | WW (5 d) | OD (105 °C/12 h) | 15 | $\sigma_c$, $E$, $V_p$ decrease | Huang et al. (2020a) |
| Sandstone | NSCB specimen | WW (25 °C/24 h) | AD (25 °C/6 d) | 50 | $P$ increase;<br>$V_p$, $K_{IC}$, decrease | Cai et al. (2020a) |
| Tuff | CS (L/D ratio of 2.5) | WW (15–24 °C/24 h) | OD (105 °C) | 52 | $A_w$ $P$ increase;<br>$\sigma_c$, $V_P$, decrease | Topal and Sözmen (2003) |
| Sandstone | CS (Φ50 × 25) | WW (25 °C/24 h) | AD (25 °C/6 d) | 50 | $A_w$ $P$ increase;<br>$V_P$, SDI,<br>$\sigma_t$ decrease | Zhou et al. (2018b) |
| Sandstone | CS (Φ50 × 25/100) | WW (25 °C/25 d) | OD (110 °C/24 h) | 5 | $\sigma_c$, $\sigma_t$, $E$, decrease | Yao et al. (2020c) |
| Granite | CS (Φ50 × 25) | WW (25 °C/10 h) | OD (50 °C/10 h) | 100 | $R$ increase;<br>$H$ decrease | Zhao et al. (2020a) |
| Sandstone | CS (Φ50 × 50) | WW (25 °C/48 h) | AD (25 °C) | 50 | $A_w$ $P$ increase;<br>$V_P$, SDI, $\sigma_{cd}$,<br>$E_d$ decrease | Zhou et al. (2018a) |
| Iron ore | CS (Φ50 × 100) | GWW (8 h)<br>WW (48 h) | AD (26 °C/7 d) | 15 | $\sigma_c$, $E$, $V_P$ decrease | Yang et al. (2018b) |
| Sandstone | CS(Φ50 × 100) | GWW (6 h)<br>WW (6 h) | OD (50 °C/12 h) | 25 | $A_w$ increase;<br>$V_p$, $\sigma_c$, $E$ decrease | An et al. (2020) |
| Sandstone | CS(Φ50 × 100)<br>CS(Φ50 × 50) | GWW (6 h)<br>WW (12 h) | OD (50 °C/12 h) AD (48 h) | 30 | $A_w$, $\varepsilon_c$ increase;<br>$\sigma_c$, $E$, $c$, $\varphi$ decrease | Li et al. (2021d) |
| Sandstone | CS (Φ50 × 25) | GWW (6 h)<br>WW (18 h) | AD (3 d) | 15 | $\sigma_t$, decrease | Zhao et al. (2017b) |
| Mudstone | CS (Φ25 × 50) | WWV (−10 MPa/24 h) | OD (60 °C/24 h) | 12 | $P$ increase<br>$\sigma_c$ decrease | Zhao et al. (2018a) |

(continued)

**Table 2.2** (continued)

| Rock type | Sample size | Treatment methods | | Cycle number | Test results | Reference |
|---|---|---|---|---|---|---|
| | | Wetting | Drying | | | |
| Sandstone | CS (Φ50 × 100) | WWV (4 h)<br>WW (44 h) | OD (45 °C/20 h); AD (4 h) | 15 | *P* increase<br>$V_P$; $\sigma_c$, *c*, decrease | Zhang et al. (2014) |
| Sandstone | CS | WWV (4 h)<br>WW (44 h) | OD (105 °C/20 h); AD (4 h) | 20 | $\sigma_c$, *E*, decrease | Xie et al. (2018a) |
| Sandstone | CS (Φ50 × 25/50) | WWV (−80 kPa/24 h) | OD (105 °C/24 h) | 10 | $A_w$ increase;<br>$V_p$, $\sigma_t$ decrease | Liu et al. (2016) |
| Sandstone | SCB specimens; CS (Φ50 × 50) | WWV (−80 MPa/24 h);<br>WW (44 h) | OD (105 °C/24 h) | 10 | $K_{IC}$, $\sigma_t$ decrease | Liang and Fu (2020) |

*CS* cylindrical sample, *OD* oven-dried, *AD* air-dried, *GWW* gradually water wetting (1/3 forward), *WW/WWV* water wetting without and with vacuum condition, *BD* Brazilian disk, *NSCB/SCB* notched/semi-circular bend, *CCBD/CCNBD/CSTBD* Brazilian disk with centrally cracked/cracked chevron notch/cracked straight-through, *R* surface roughness, *Tc* thermal conductivity, $E_d$ dynamic Young's modulus, $\sigma_{cd}$ dynamic compressive strength, $\sigma_t$ tensile strength, $K_{eff}$ effective fracture toughness, $K_{IIC}$ mode II fracture toughness, $K_{IC}$ mode I fracture toughness, *SDI* slake durability index, *PLI* point load index, *P* porosity, $A_w$ water absorption rocks (especially those with strong hydrophilicity, such as mudstone) disintegrate after immersion for a short period (Azhar et al., 2020; Fujii et al., 2020). In addition, many studies have shown that some rocks do not weaken even after long-term immersion (Ai et al., 2021).

As the immersion time increases, the mechanical parameters ($\sigma_c$, $\sigma_t$, $E$, $c$, $\varphi$, $\tau$) of some rocks are weakened, their brittleness decreases, the failure mode becomes stable, and the roughness of the fracture planes increases (Zhu et al., 2019; Ma et al., 2021). The time-dependent immersion weakening varies for different rocks. For example, as the immersion time increases, the uniaxial compressive strength of coarse sandstones first decreases rapidly, then increases slightly, and finally decreases (Wu et al., 2020a). The uniaxial compressive strength and elastic modulus of argillaceous slates decrease with increasing immersion time, whereas Poisson's ratio remains roughly unchanged (Huang et al., 2020b).

### 2.2.4 *Other WRI-Related Factors*

The chemical composition and water pressure play significant roles in WRI weakening. The mechanical parameters ($\sigma_c$, $\sigma_t$, $E$, $c$, $\varphi$, $\tau$) of chemically treated samples generally demonstrate more significant weakening compared with natural immersion conditions, especially for pre-fractured samples (Zhang et al., 2019g; Gong et al., 2021). For example, the salts contained in water can gradually accumulate in the pore networks of rocks under wetting-drying cycles, which can cause rock deterioration (Jiang et al., 2022a). The water immersion height also influences rock strength; the strength of partially presoaked specimens is lower than that of wholly presoaked specimens (Chen et al., 2021a). Seepage water pressure enhances the deformation resistance of rock and affects rock strength. As seepage pressure increases, the stress thresholds for crack initiation and damage during rock compression decrease (Xiao et al., 2020; Zhong et al., 2019; Li et al., 2020b). Generally, studies on the WRI of rock mass cover various factors, and strength tests investigate the weakening characteristics under various immersion conditions to reflect engineering environments. Then, it is used to predict the degree of influence of the WRI on engineering scales.

## 2.3 Weakening Mechanism and Microscopic Characterization

Many studies have shown that WRI is mainly the interaction between water and clay-related minerals in rocks, which changes the pore structure and further degrades their strength. Therefore, in addition to characterizing the macroscopic strength, investigating the internal microstructure is a valuable way to uncover WRI mechanisms. Microscopic observations include laser scanning confocal microscopy

(LSCM) (An et al., 2020), polarizing microscopy (PM), scanning electron microscopy (SEM) (Dehestani et al., 2020; Zhang et al., 2014; Liu et al., 2018b; Zhou et al., 2018b; Du et al., 2019a), neutron radiography (NR), nuclear magnetic resonance (NMR) (Xie et al., 2018b; Zhao et al., 2017a, 2018a, 2018c, 2019b), computed tomography (CT), small angle X-ray scattering (SAXS) and other methods (Liu et al., 2016; Zhao et al., 2014, 2019a; Wang et al., 2021a). The main application of the micro-observation techniques is shown in Fig. 2.2. The XRD pattern shown in Fig. 2.2 is typically used to investigate the hydrophilic mineral composition of the rocks.

The microinvestigation methods shown in Fig. 2.2 can be divided into three categories: SEM, PM, and LSCM. They are primarily used to observe the surface structure of a rock mass. SEM can be used to observe the mineral occurrence

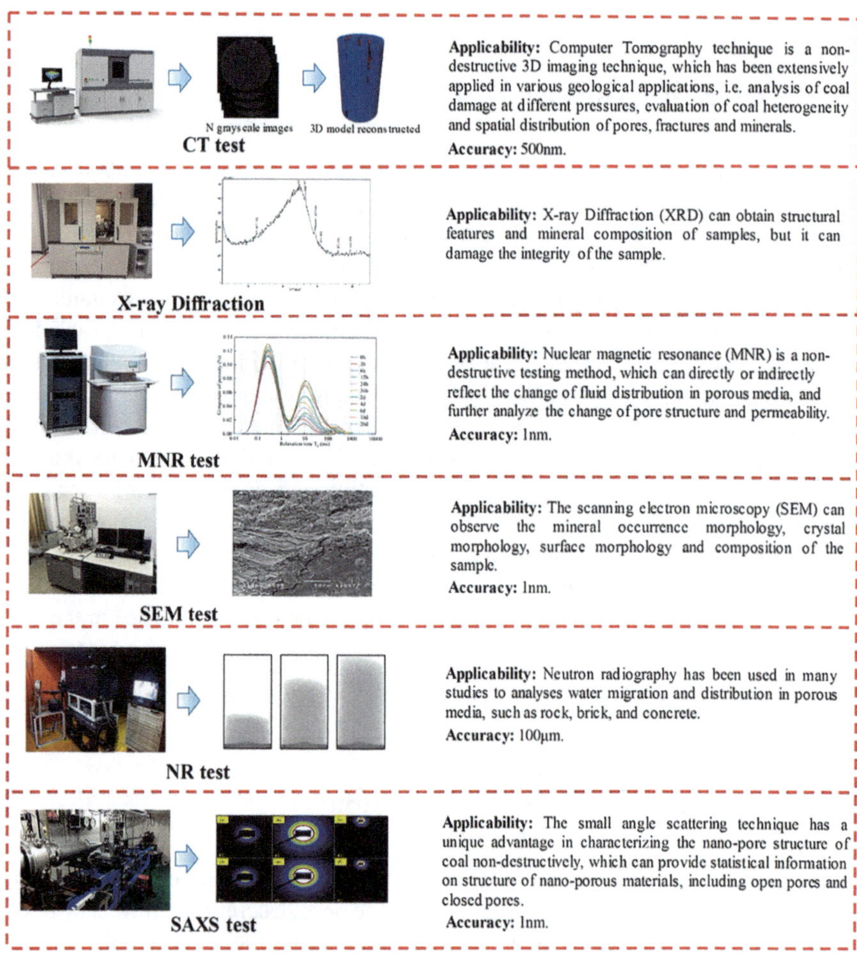


**Fig. 2.2** Main characterization means of pore structure in WRI analysis

morphology, crystal morphology, surface morphology, and composition. However, the tested samples must be sprayed before SEM scanning; therefore, it is a destructive test. LSCM is mainly used for scanning the fracture surface. MNR and NR are mainly employed to judge pore structures and immersed liquids. The MNR is a non-destructive test that is widely used to characterize pore structures. However, NMR cannot be used to reconstruct a three-dimensional pore structure. CT, SAXS, and XRD are used for X-ray fluoroscopy, and the pore structure and mineral composition can be reconstructed with post-processing. CT is a non-destructive technique that can reconstruct pore and fracture structures in real time and is widely used to track pore and fracture evolutions during rock deformation. However, in the process of three-dimensional reconstruction of CT images, the division of pore fracture and mineral composition threshold is generally manually defined, which sometimes affects the accuracy of the reconstruction model.

Thus, various microcharacterization techniques are generally used for complementary analysis (Zhang et al., 2021f; Ai et al., 2021). For example, NMR, XRD, and CT can be used to accurately reconstruct pore structures and mineral compositions (Fig. 2.3). Specifically, the mineral composition in the sample is first identified by XRD, such as clay mineral composition with strong hydrophilicity, and then the threshold is used during the reconstruction of CT images. Similarly, NMR can accurately provide the pore structure division threshold for CT reconstruction.

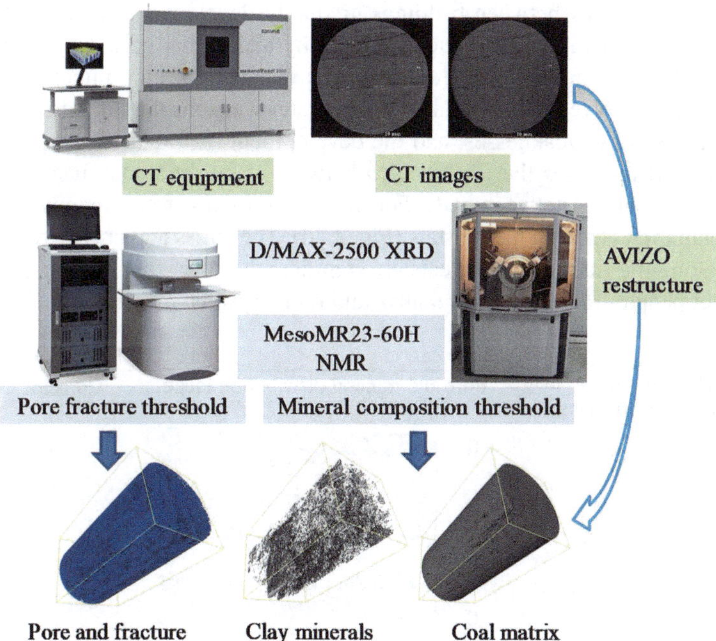


**Fig. 2.3** Reconstruction process of "pore-fracture" dual structure of a coal sample

Based on the CT reconstruction method (Fig. 2.3), we reconstructed the pore structure and mineral composition model of the coal samples before and after water immersion. Table 2.3 shows the distribution of the pores with increased connectivity (blue) and reduced clay minerals (red) before and after immersion. The locations where the clay minerals are reduced coincide with the positions where the connectivity pores are increased. Further, the change in the pore structure in the sample during WRI is mainly caused by the dissolution and expansion of hydrophilic mineral components (Azhar et al., 2020; Huang et al., 2020b; Liu et al., 2021).

Currently, five mechanisms for rock strength weakening caused by water immersion have been proposed: (1) expansion and dissolution of clay minerals, (2) reduction of capillary tension, (3) increase in pore pressure, (4) reduction of fracture energy, and (5) weakening of intergranular cohesion and friction (Zhu et al., 2020a; Li et al., 2020a). By comparing the CT images before and after immersion (Fig. 2.4), the internal damage process caused by the WRI can be observed. Figure 2.4a, b show the CT images at the same position before and after immersion. The darkening of the grayscales in the CT images indicates an increase in damage and a decrease in density. Therefore, pre-existing weaknesses are displayed in darker colors. As shown in Fig. 2.4a, there are apparent darker parts (weaknesses) in the dry sample because coal is a typical porous medium with a large number of pore structures (connected pores and isolated pores) and fracture structures (Yao et al., 2019a). After immersion, the size and aperture of the internal fractures in the coal sample increased, and cleats between beddings gradually developed with water degradation. Fracture structures have better connectivity and higher permeability, forming the main flow channels of the fluid medium; meanwhile, clay minerals near the fractures are dissolved in water. In addition, water weakens the intergranular bonding of pre-existing weaknesses, and the development of fractures and pores (connected pores) increases the contact area between water and coal, resulting in the appearance of cleats between beds. For the rock samples, water mainly weakened the cementation between the crystals (Fig. 2.4d). The stress and energy required for failure along the fractures and cleats are significantly lower than those required for the direct penetration of the coal matrix and rock crystals; therefore, the higher the

**Table 2.3** Distribution map of pore and mineral changes

| Pore and mineral distribution location | Coincidence ratio |
|---|---|
| a | 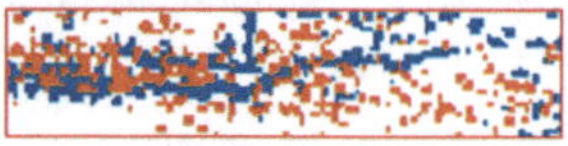 59.82% |
| b | 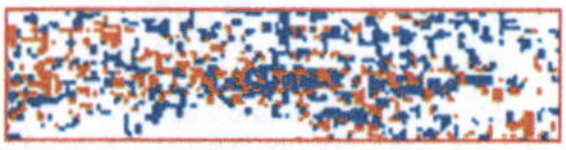 63.14% |
| c | 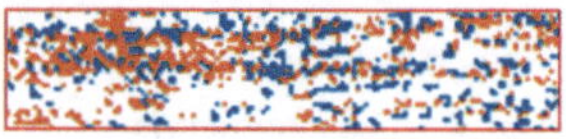 65.38% |
| d | 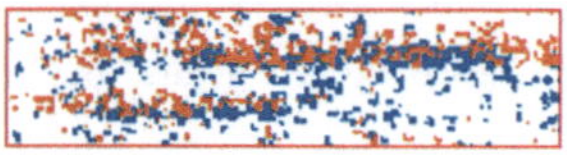 61.97% |

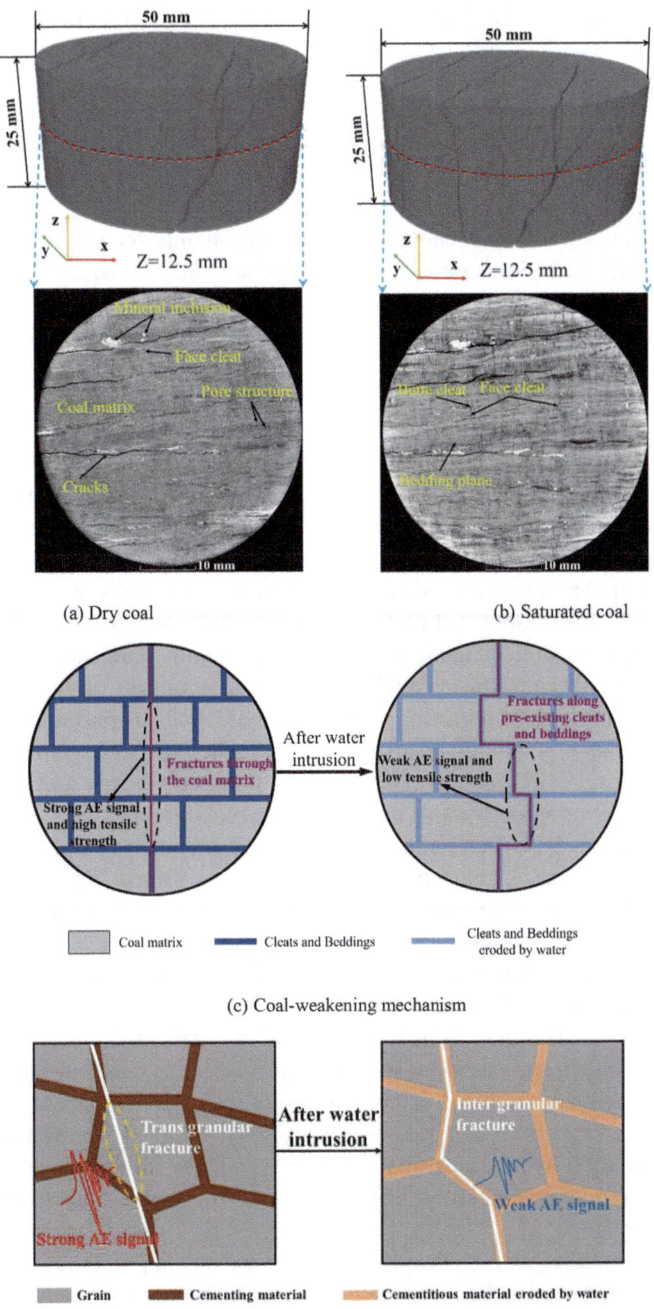


**Fig. 2.4** CT images of a coal sample before and after water immersion. (**a**) Dry sample, (**b**) Saturated sample, (**c**) Coal weakening mechanism, and (**d**) Rock weakening mechanism

water content, the smaller the rock strength and the weaker the AE signal (Fig. 2.4) (Li et al., 2021a, 2021b; Miao et al., 2021).

## 2.4 Conclusion and Prospect

Research on the water weakening of rock masses mainly focuses on the effects of water content, water immersion time, and cyclic water immersion. The strength-weakening degree is characterized by uniaxial and triaxial strength, shear strength, tensile strength, point load, penetration, etc. AE and IR methods are typically used to study the fracture and energy characteristics during loading under WRI. The strength of the rock mass decreases to varying degrees with water content, immersion time, and WD cycles and is related to the type of rock mass and mineral composition. Generally, the strength decreases exponentially with an increase in water content. The previous several WD cycles have a significant impact on the strength of the rock mass and have little effect on the progression of WD cycles. The degree of rock weakening gradually decreases with an increase in the immersion time. The corresponding AE count, intensity, and IR intensity also weaken accordingly. The WRI enhances the plasticity of rocks and reduces their brittleness.

Various microscopic methods have been used to study pore characterization and the weakening mechanism of WRI. SEM can qualitatively observe the fracture structure and mineral composition of the rock mass, but it is impossible to directly compare the porosity before and after water immersion owing to gold spraying during observation. NMR can quantitatively determine the porosity and pore size distribution and is a non-destructive test technique, making it possible to compare the pore size changes before and after immersion. CT scanning combined with corresponding reconstruction algorithms can quantitatively compare the changes in pore structure and mineral composition before and after immersion. However, the threshold division in CT image reconstruction is significant and directly affects the accuracy of the analysis. Thus, various micro-methods can be used to study the evolution of pore structure changes under WRI.

In the WRI weakening mechanism, the clay minerals in the rock mass dissolve in water, which expands the pore structures, increases the connected pores, further expands the primary fractures, and consequently increases the porosity and permeability. At the same time, the presence of water weakens the cementation strength near the primary fractures, making it easier to expand the splitting fracture along the joint surfaces. All these weakening processes lead to a decrease in rock strength and AE intensity.

Currently, studies on both the physical and mechanical properties and the weakening mechanism of WRI are relatively mature. Targeted experiments can only be carried out for special rock masses (different mineral compositions) or water environments (different water chemical compositions) in on-site engineering (water influence conditions). On this basis, the corresponding mechanical test modes (tension, compression, shear, or other tests), water influencing factors (moisture content, wetting-drying cycles, immersion time, etc.) and macro (AE, AIRT, etc.) and micro (SEM, NMR, CT, etc.) characterization methods can be considered. Thus, more attention should be paid to how laboratory-scale tests can be applied in engineering-scale practice. Therefore, some qualitative conclusions must be transformed into quantitative models, which can then be applied to field engineering problems through numerical simulations. In addition, when investigating the effect of the immersion time, most laboratory tests are applied for less than half a year; however, long-term water immersion (even hundreds of years) problems generally occur. The weakening mechanism of WRI in rocks can also be explained using the changes in water immersion chemical ions and components.

# Chapter 3
# Accurate Characterization Method of Pores and Various Minerals

## 3.1 Introduction

To study the relationship between the internal pore structure and mineral composition of rock mass and its physical mechanics and seepage behavior, previous studies have conducted a lot of research in pore structure and mineral composition characterization. With the progress and development of testing equipment, mercury intrusion method (Turturro et al., 2022), nuclear magnetic resonance (NMR) (Cai et al., 2018; Feng et al., 2022), scanning electron microscope (Ma et al., 2022a), acoustic emission (Yuan et al., 2023), etc. have been gradually applied to the study of microstructure. Mercury intrusion method has a large span of pore scale. It is widely used to test the pore volume, pore size distribution, and connectivity of large and medium pores in coal. However, this method will destroy the sample and seriously underestimate the content of macropores. Because the pressure it measures reflects the diameter of the pore entrance, not the diameter of the pore itself (Tang et al., 2021b; Zhu et al., 2016). The scanning electron microscope can observe the occurrence form, crystal form, surface morphology, and composition of minerals. But the same method will destroy the test sample, and the observation range is small. And it is difficult to provide detailed information on the porosity and pore connectivity of large orebodies (Dehestani et al., 2020; Zhang et al., 2014; Du et al., 2019b). As a non-destructive testing method, NMR is widely used to characterize the pore structure. Although the measurement accuracy is high, it can neither reconstruct the three-dimensional pore structure, nor can it further obtain other mineral information of coal and rock mass (Luo et al., 2022; Zhao et al., 2019d; Kuang et al., 2023). In recent years, X-ray computed tomography (CT) has proven useful in 3D imaging and characterizing internal rock properties, such as pore networks, fracture geometries, and heterogeneous minerals (Gupta et al., 2022). The real-time computed tomography system and image processing technology equipped with loading equipment provide an effective way to further quantitatively study the failure patterns of the inner space of rocks under compressive loads (Fan et al., 2022; Yang et al.,



C. Zhang, *Water Rock Interaction in Underground Coal Mining*,
https://doi.org/10.1007/978-981-95-9957-8_3

2022). Real-time CT can visually present the spatial three-dimensional fracture shape and two-dimensional local fracture information of the rock after compression failure. The detection accuracy depends on the resolution of the CT scan and the sample size (Li et al., 2022a, 2022b) because CT has the advantages of being non-destructive, dynamic, and continuous. Combined with specific reconstruction software, quantitative characterization and three-dimensional reconstruction of coal and rock mass pores and mineral components can be performed (Gao et al., 2022; Zhang et al., 2022f). However, the threshold division of porosity and mineral composition in the process of CT image characterization determines the reliability of the characterization (Wang et al., 2023b). Table 3.1 shows the relevant research using CT scanning to analyze the pores and mineral components of coal and rock mass.

It can be seen from Table 3.1 that it has become a useful method to characterize and reconstruct the pore structure and mineral composition of coal and rock mass by CT scanning. However, the division of the pore threshold in the current research is mainly based on image processing technology and direct human judgment. Comparative analysis of multiple means is rarely performed. Meanwhile, for the mineral composition, all minerals are grouped together. There is no threshold for distinguishing various minerals in coal and rock mass. This is not conducive to analysis of changes in mineral composition in coal and rock mass. For example, under the action of water-rock interaction, different mineral components interact with water in different forms (Zhang et al., 2023b; Han et al., 2023). If various minerals can be accurately divided, the mechanism of pore evolution before and after water immersion can be analyzed. Accordingly, a method to determine the threshold value of various mineral components and pores in CT reconstruction was proposed. Based on this method, threshold segmentation and three-dimensional reconstruction of coal samples were carried out. This method can be used as an important means for microscopic analysis of coal and rocks.

## 3.2 Test Samples and CT Scans

### *3.2.1 Test Sample*

The coal samples studied in this chapter were taken from the −710 m mining level, 3# coal seam of Datai Coal Mine. The location of the studied coal mine is shown in Fig. 3.1. This coal seam is a steeply inclined coal seam with an average buried depth of 820 m and an average angle of 45°–88°. The scanning electron microscope was used to observe the microstructure of the coal sample, as shown in Fig. 3.2.

Figure 3.2 shows that the coal sample is mainly composed of coal matrix, minerals, and pores and fractures. The internal structure is relatively complex, and the pores and fractures are also divided into connected fractures and isolated pores. Most of the minerals are present in the coal matrix. Moreover, it can be clearly seen that there are more connected pores in the coal sample, and less isolated pores. Connected pores are usually larger in diameter and thin strips, and isolated pores are smaller.

**Table 3.1** Case study of CT scanning characterization of pores and mineral composition

| No | Sample | Pore threshold determination | Mineral threshold determination | Method | Reconfiguration software | Research content | References |
|---|---|---|---|---|---|---|---|
| 1 | Coal | Visual recognition | / | Build a Bi-PTI model | Matlab, Avizo | Pore structure characterization | Hao et al. (2022) |
| 2 | Coal | Visual recognition | / | / | Datosx, Vgstudio | Characterizing coal fractures under compressive loads | Li et al. (2020c) |
| 3 | Coal | Visual recognition | Visual recognition | / | Mimics | 3D characterization of minerals, pores, and fractures in coal | Yao et al. (2009) |
| 4 | Coal | Visual recognition | Visual recognition | Otsu | Matlab, Avizo, Geomagic Studio, Hypermesh | 3D characterization of minerals, pores, and organic matter | Zhao et al. (2018) |
| 5 | Coal | Visual recognition | Visual recognition | / | Xrm Reconstructor, Avizo | 3D characterization of minerals, pores and fractures | Shi et al. (2018) |
| 6 | Coal | Visual recognition | Visual recognition | / | / | 3D characterization of pore and fracture structures due to diffusion deposition of grouting | Tao et al. (2021) |
| 7 | Coal | Visual recognition | Visual recognition | Otsu | Mimics | 3D characterization of pore and fracture structures under compressive loads | Du et al. (2022a) |
| 8 | Coal and rock | Visual recognition | Visual recognition | Otsu | Mimics | 3D characterization of pore and fracture structures under compressive loads | Du et al. (2021) |
| 9 | Coal and rock | / | Visual recognition | Otsu | Matlab, Avizo, Mimics | 3D characterization of mineral structures | Fu et al. (2020b) |
| 10 | Coal | Visual recognition | Visual recognition | Otsu | VG Studio MAX, ImageJ, AVIZO | 3D characterization of pore and mineral structures before and after loading | Fu et al. (2020a) |

(continued)

**Table 3.1** (continued)

| No | Sample | Pore threshold determination | Mineral threshold determination | Method | Reconfiguration software | Research content | References |
|---|---|---|---|---|---|---|---|
| 11 | Coal and rock | Visual recognition | / | Otsu | Matlab, Photoshop, Mimics | 3D characterization of pore and fracture structures before and after loading | Li et al. (2015) |
| 12 | Coal | Visual recognition | / | Otsu | Matlab, VGStudio MAX | 3D characterization of pore and fracture structures of gas-bearing coal under triaxial compression loading | Wu et al. (2022) |
| 13 | Coal | Comparison with NMR measurements | Comparison with XRD measurements | / | Avizo | 3D characterization of pore and mineral structures before and after water immersion | Zhang et al. (2023a) |
| 14 | Concrete | Visual recognition | / | / | 3D Structured-Light Scanner | 2D characterization of the fractures under cyclic loading | Deng et al. (2023) |
| 15 | Limestone | Visual recognition | / | / | / | 3D characterization of pore and fracture structure after different freeze-thaw times | Li et al. (2022a, 2022b) |
| 16 | Coal | Visual recognition | / | Otsu | Phoenix Datos\|X2, Avizo | 3D characterization of pore and fracture structure | Mao et al. (2022) |
| 17 | Coal and rock | Visual recognition | Visual recognition | Otsu | Avizo | 3D characterization of pore and mineral structures | Chen et al. (2022b) |
| 18 | Coal gangue | Visual recognition | / | Otsu | Avizo | 3D characterization of pore structure of crushed coal gangue under load compression | Li et al. (2022a, 2022b) |
| 19 | Sandstone | Visual recognition | / | / | Avizo | 3D characterization of pore and fracture structure | Yu et al. (2021) |
| 20 | Coal | Comparison with porosity measured by MIP | / | / | Scanip, Avizo | 3D characterization of pore and fracture structure | Wang et al. (2021b) |
| 22 | Ore | / | Visual recognition | Otsu | Ors Dragonfly | 2D characterization of the mineral composition | Guntoro et al. (2021) |

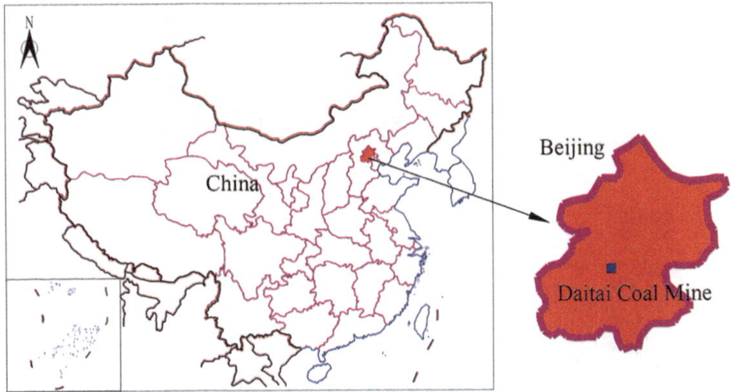


**Fig. 3.1** The location of the studied coal mine

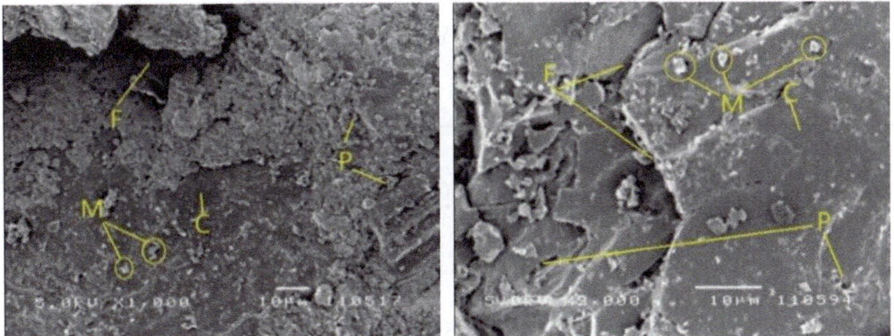


C—coal matrix, F—connected pores, M—Minerals, P—isolated pores

**Fig. 3.2** Distribution of minerals, pores and fractures and coal matrix in coal samples. C—coal matrix, F—connected pores, M—minerals, P—isolated pores

### *3.2.2 CT Scan*

CT scanning technology is a high-resolution non-destructive testing method. An industrial X-CT system usually consists of an X-ray source, a sample platform, a radiation detector, and a computer system to analyze the data (Dong & Qiao, 2021; Gupta et al., 2022). The ability of X-rays to penetrate the sample is inversely proportional to the density of the sample itself. On CT grayscale images, low-density regions such as voids or fractures appear darker, whereas high-density regions such as minerals appear gray or brighter (Li et al., 2022a, 2022b).

This test used nanoVoxel-3000 CT scanning equipment from Tianjin Sanying Technology Co., Ltd., as shown in Fig. 3.3. The scanning voltage was 160 kV, the scanning current was 175 μA, the number of scanning frames was 1800 frames, and the scanning accuracy was 20 μm. One CT scan was performed on the sample, and a three-dimensional grayscale matrix of 1200 × 1200 × 700 pixels was

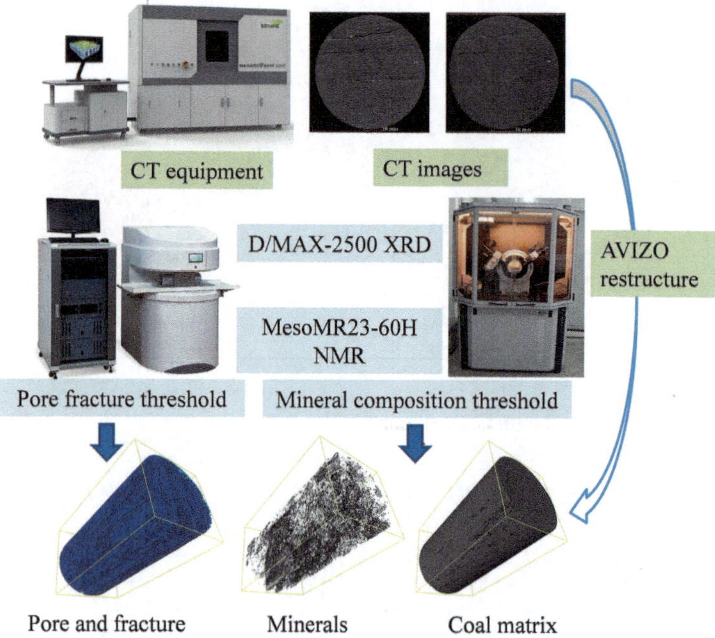


**Fig. 3.3** Threshold segmentation principle in 3D reconstruction

obtained each time. In order to facilitate storage and follow-up experiments, the three-dimensional grayscale matrix was cut into 700 grayscale images with a size of 1200 × 1200 pixels along the height direction of the disc, each with a thickness of 1 voxel. Common CT images are 8-bit or 16-bit grayscale images, and 16-bit grayscale images contain more grayscale information and are used in this chapter (Mao et al., 2022).

## 3.3 Threshold Segmentation for Pores and Minerals

### *3.3.1 Model 3D Visualization*

At present, there are many numerical analysis software that can visualize and further analyze the grayscale images obtained by CT scanning, such as Thermo Scientific Avizo software, VGStudio software, MATLAB software, COMSOL software, and MIMICS software. In this chapter, Avizo software is used to superimpose and three-dimensionally display the grayscale images obtained by CT scanning (Fig. 3.4). And then the threshold segmentation and three-dimensional reconstruction are carried out to further quantitatively analyze the coal samples.

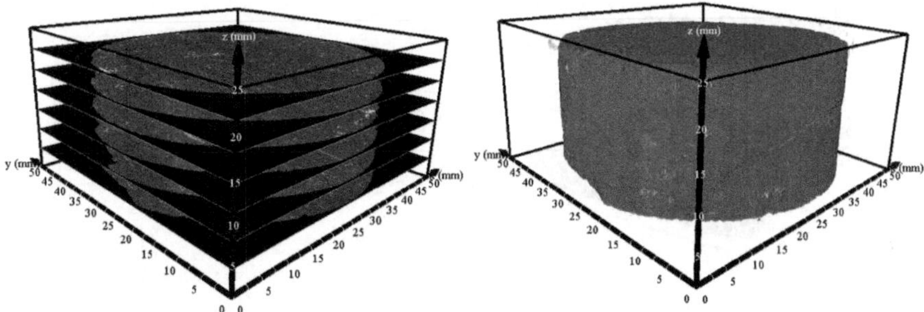

Fig. 3.4 3D visualization for CT scan images

### 3.3.2 Select the Analysis Area

If the raw coal sample is directly quantified and analyzed, the result will have large errors. There are two main reasons: (a) When coal samples are processed on site, stress concentration will be generated externally due to machine cutting. This may lead to more obvious secondary fractures on the outside of the coal sample. (b) Image artifacts caused by inherent physical limitations of X-ray transmission technology during CT scanning of coal samples. This phenomenon is more serious at both ends of the coal sample (Godinho et al., 2021; Gupta et al., 2022; Heyn et al., 2021; Kamel et al., 2022; Wang et al., 2021a). As shown in Fig. 3.5a, they are the grayscale images obtained by the 50th and 650th CT scans, respectively. Therefore, in order to reduce the error of the data and improve the accuracy of the analysis results. It is necessary to select appropriate regions within the coal sample for analysis.

### 3.3.3 Threshold Segmentation

The threshold range of pore and fracture structure and mineral composition in coal samples affects the accuracy of 3D reconstruction (Wang et al., 2023a). Therefore, the content of mineral components is measured by an X-ray diffractometer in this chapter. During the threshold division, the threshold value is constantly adjusted to make the content of mineral components after the division close to the results measured by the diffractometer. In this way, the threshold range of mineral composition is obtained. The internal porosity of coal samples can be measured by nuclear magnetic resonance. When the size of the porosity is known, the threshold range of the porosity is obtained in the same way, as shown in Fig. 3.3.

(a) Representative elementary volume determination

To obtain the representative elementary volume for coal sample reconstruction, a total of 28 cube regions of different sizes were cut from the center of the

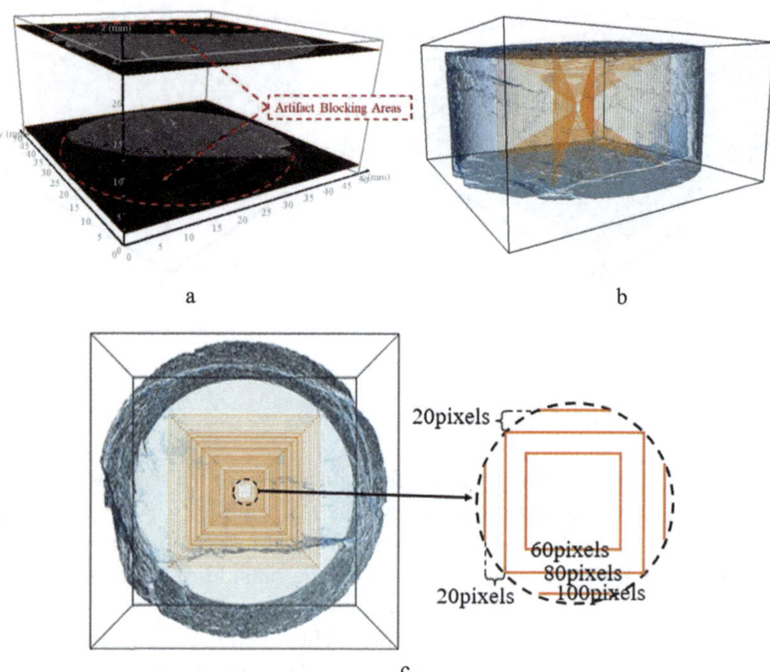


**Fig. 3.5** Select the analysis area (**a**. is the 50th and 650th grayscale images; **b**. and **c**. are perspective views of the cut area)

coal sample (Fig. 3.5c). The length of the cut area is increased by 20 pixels from inside to outside, and the side length is at least 60 pixels and at most 600 pixels, as shown in Figures 3.4c and 3.5b. In this chapter, MesoMR23-060H medium-sized nuclear magnetic resonance imaging analyzer was used to measure the porosity of the coal sample, as shown in Fig. 3.3. In this chapter, an NMR test is used to determine the pore threshold in CT reconstruction. NMR techniques can test the pore structure and fluid filling characteristics of 1H-bearing fluids in rock-like materials based on their relaxation times, T2. In particular, the peak area of the T2 atlas is related to the number of pores within the material, and the peak position is related to the pore size. According to previous research, for coal samples, the relaxation times, T2, being 0.5–2.5 ms, 20–50 ms, and > 100 ms correspond to tiny pore (<0.1 μm), medium-to-large pore (0.1~1 μm), and crack (>1 μm) information, respectively (Yu et al., 2022). However, the accuracy of CT scanning in this chapter is 0.1 μm. Thus, the pores mentioned in this chapter are mainly medium-to-large pore (0.1~1 μm) and crack (>1 μm). Considering the accuracy of CT scanning, the threshold of pores in the CT reconstruction is only selected for the parts larger than 0.1 μm in NMR measurement. The porosity of the coal sample was measured to be 6.88% according to the T2 spectrum. By constantly adjusting the size of the threshold, the pore content is the same as the result of NMR. The resulting thresholds were recorded and plotted in

Fig. 3.6a. As can be seen from Fig. 3.5a, when the side length of the cube is greater than 140 pixels (5 mm) and less than 400 pixels (13.6 mm), the threshold fluctuation is small. The XRD test of the natural coal sample was carried out by using the D/MAX-2500/PC model equipment (Fig. 3.3). The mineral content in the coal sample was measured to be 11.2%. Similar to the characterization of pores, the thresholds corresponding to different sizes for minerals are shown in Fig. 3.6b. It can be seen that the representative elementary volume of mineral reconstruction has a side length of 140 pixels. The threshold values remain stable until the side length exceeds 580 pixels (20.7 mm). Therefore, to ensure the reliability of the analysis, the subsequent analysis mainly focuses on the cube with a side length of 300 pixels. The pore threshold is determined to be 19,840, and the threshold for the mineral composition is 24,533.

For comparison and verification, three suitable areas will be selected for analysis in the entire coal sample, as shown in Fig. 3.6c. The relative coordinates of region A from the origin are 173 pixels, 186 pixels, and 125 pixels. The relative coordinates of area B from the origin are 612 pixels, 253 pixels, and 231 pixels. The relative coordinates of area C from the origin are 274 pixels, 672 pixels, and 283 pixels. The pores and minerals of regions A, B, and C are divided using determined thresholds. The porosity of these three cubes was 6.52%, 6.72%, and 7.04%, respectively. The result is close to that of NMR measurement, indicating the determined pore

(a) Pores (b) Minerals

(c) Study area location

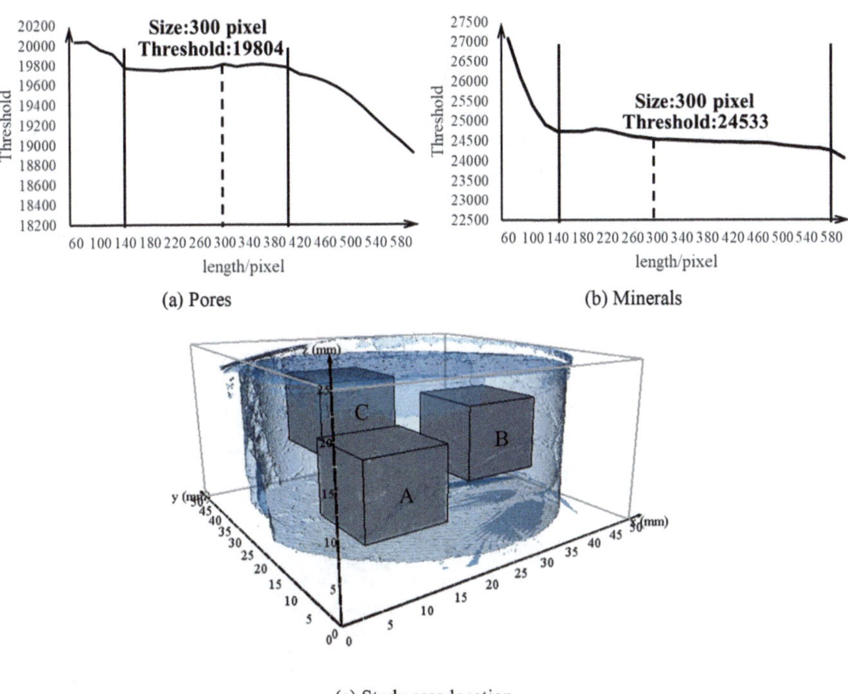

**Fig. 3.6** Identifies the analysis area. (**a**) Pores, (**b**) Minerals, (**c**) Study area location

threshold is more accurate. The content of mineral components was calculated as 11.1%, 11.5%, and 11.7%, respectively. The deviation is less than 5%, indicating that the threshold value of mineral composition is reliable.

## 3.4 3D Reconstruction for Pores and Minerals

After the threshold treatment of mineral composition and pores, noise reduction is also required to reduce the errors caused by the defects of CT scanning technology. Then, 3D reconstruction and quantitative analysis were performed for each segment component. As shown in Fig. 3.7, the blue area indicates pores, and the yellow area indicates minerals. In addition to mineral components and pores, the other main groups of coal samples are divided into coal matrix. After the entire coal sample is reconstructed, pores and minerals are subtracted from the entire model. The rest of the model is the coal matrix. The gray area in Fig. 3.7 represents the coal matrix in region A.

### *3.4.1 Division of Pores*

Pores in coal samples can be divided into connected pores and isolated pores according to connectivity. The connected pores have a great influence on the mechanical properties and permeability of coal samples. The connected pores in the pore group were extracted by the Axis connectivity command included in the software. Then the Arithmetic command subtracted the connected pores from the pore group to get the isolated pores. Figure 3.8 shows 3D display models of connected and isolated pores in regions A, B, and C, respectively. The proportion of connected pores in

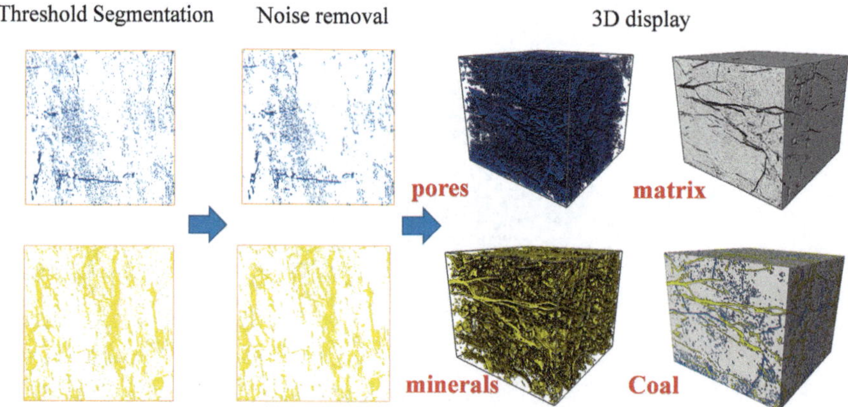


**Fig. 3.7** 3D reconstruction of coal sample

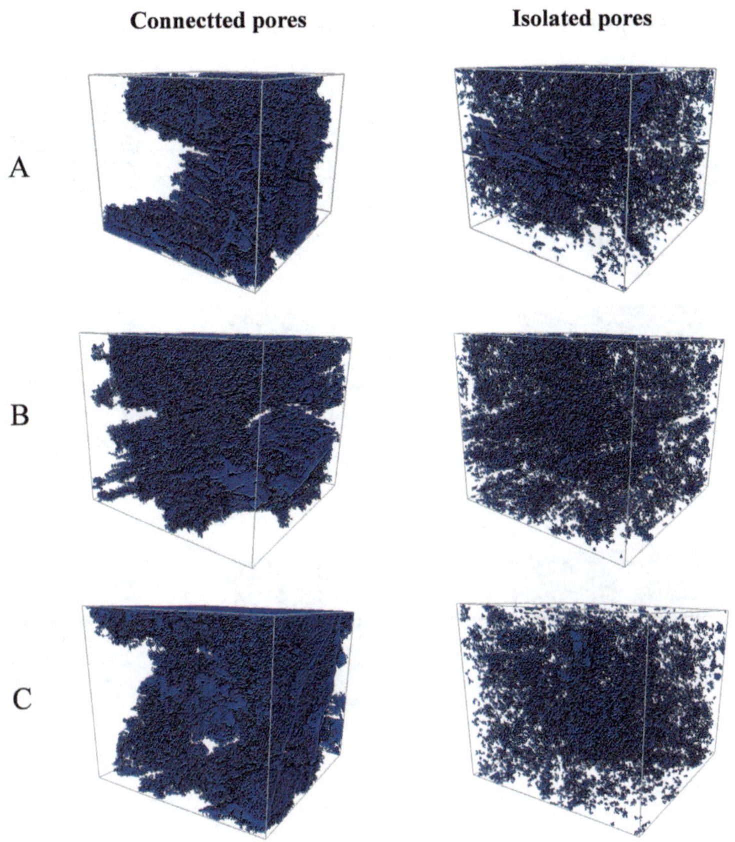


**Fig. 3.8** Classification of coal sample pores

regions A, B, and C is 5.34%, 5.17%, and 5.81%, respectively. The isolated pores were 1.54%, 1.55%, and 1.23%, respectively.

The label analysis command was used to classify and quantify the pores. Figure 3.9 shows the 3D model of pore distribution in regions A, B, and C, respectively. Different colors correspond to different pores. Because there are fewer colors than the number of pores, some pores are labeled with the same color. The calculated pore parameters of regions A, B, and C are shown in Table 3.2. There are 9894, 10,605, and 8055 pores in coal samples in regions A, B, and C, respectively. Although there is only one (region A and C) or two (region B) connected pores, the volume is much larger than any single pore. This is also the reason why the volume proportion of connected pores is greater than that of isolated pores. In addition to volume, equivalent diameters and form factors of connected pores are also much larger than those of isolated pores in regions A, B, and C. The equivalent diameter indirectly reflects the seepage capacity of the pore. The larger the equivalent diameter of the pore, the stronger the seepage capacity. The shape factor usually indicates pore shape complexity, irregularity, or discontinuity. The shape factor of the

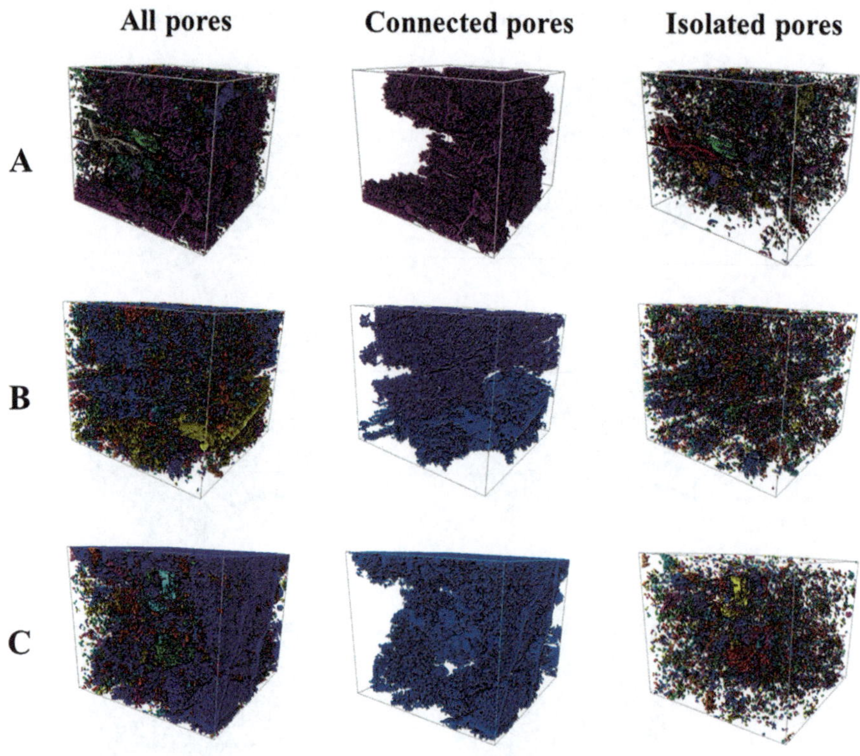


**Fig. 3.9** Distribution of pore size

connected pore is large, indicating that it contains many small pores and channel twists. Flatness represents the ratio of the maximum transverse size to the maximum longitudinal size of the pore. When the ratio is close to 1, the pore shape is more spherical or square. When the ratio is close to 0, it indicates that the pore shape is closer to the needle shape (Ma et al., 2022b; Fusi et al., 2013; He et al., 2018; Zhang et al., 2016d). It can be seen from the table that the flatness of the connected pores in each region is greater than 0.5, indicating that their shape is close to round or square. The distribution of the isolated pores flatness ratio ranges from 0 to 0.992, indicating that the shapes of different isolated pores vary greatly. The mean flatness is less than 0.5, indicating that the shape of isolated pores is more like a needle.

### *3.4.2 Division of Mineral Components*

From the XRD test results, the internal minerals of the coal sample are composed of various mineral components. The contents of various mineral components are shown in Tables 3.3 and 3.4. They are the average of three tests taken from different positions of coal samples, and the three test results are basically consistent. And the

**Table 3.2** Pore parameters

| Region | Pore | Average equivalent diameter | Average volume | Average shape factor | Average flatness | Maximum flatness | Minimum flatness |
|---|---|---|---|---|---|---|---|
| A | Total porosity | 0.14317514 | 0.011645022 | 16.4549026 | 0.401838511 | 0.99143672 | 0 |
| | Connected pores | 5.54942846 | 89.4834671 | 111606.188 | 0.674507499 | – | – |
| | Isolated pores | 0.14262867 | 0.00260081 | 5.17531109 | 0.401810974 | 0.99143672 | 0 |
| B | Total porosity | 0.14231504 | 0.01060288 | 16.7996864 | 0.395360619 | 0.993932843 | 0 |
| | Connected pores | 4.34304333 | 43.2383881 | 62219.7734 | 0.58675544 | 0.640049142 | 0.533461738 |
| | Isolated pores | 0.14152263 | 0.00244879 | 5.06657219 | 0.395371675 | 0.993932843 | 0 |
| C | Total porosity | 0.14220075 | 0.01462477 | 13.2286978 | 0.397181422 | 0.977143049 | 0 |
| | Connected pores | 5.70323086 | 97.1316757 | 66214.4531 | 0.58997798 | – | – |
| | Isolated pores | 0.14151021 | 0.00256613 | 5.00903654 | 0.397157431 | 0.977143049 | 0 |

**Table 3.3** X-ray diffraction analysis report of whole-rock minerals

| | Mineral content(%) | | | | |
|---|---|---|---|---|---|
| Sample | Quartz | Feldspar | Calcite | Amorphous | Clay minerals |
| Coal | 5.0 | 8.6 | 0.4 | 30.0 | 56.0 |

**Table 3.4** X-ray diffraction analysis report of clay minerals

| | Relative content of clay minerals(%) | | | | | | Mixing ratio (%S) | |
|---|---|---|---|---|---|---|---|---|
| Sample number | S | I/S | It | Kao | C | C/S | I/S | C/S |
| Coal | / | / | 47 | / | 53 | / | / | / |

*S* smectite, *I/S* Aemon mixed layer, *It* illite, *K* kaolinite, *C* chlorite, *C/S* green smudge layer

**Table 3.5** Mineral densities in coal samples

| Mineral | Quartz | Feldspar | Amorphous | Illite | Chlorite |
|---|---|---|---|---|---|
| Density g/cm$^3$ | 2.60~2.65 | 2.54~2.61 | – | 2.64~2.69 | 2.60~2.69 |
| Average g/cm$^3$ | 2.63 | 2.58 | – | 2.67 | 2.78 |
| A: Content/% | 5.0 | 8.6 | 30.4 | 26 | 30 |
| B: Content/% | 5.6 | 7.4 | 31 | 24.8 | 31.2 |
| C: Content/% | 5.2 | 8.1 | 30.8 | 25.5 | 30.4 |
| Threshold | 25,927 | 25,740 | 24,533 | 26,826 | 28,748 |

"K-value method" is used in X-ray diffraction quantitative analysis (Yuan et al., 2011). Among them, the content of clay minerals is relatively high, accounting for 56%. And 47% of the clay minerals in natural coal samples are illite and 53% are chlorite.

Due to X-rays, CT has different penetration capabilities for substances of different densities. The higher the density, the poorer the penetration ability. The densities and contents of various minerals are shown in Table 3.5. Although the densities of various minerals partially overlap, they can be distinguished by using the average density. Because the calcite content is too low, it is ignored in the division. The various minerals that can be obtained in order of density from small to large are amorphous, feldspar, quartz, illite, and chlorite. By adjusting the threshold of region A, the contents of various minerals are consistent with XRD results. After adjusting and determining the gray value range of various minerals in region A, three-dimensional reconstruction is carried out. The refactoring results are shown in Fig. 3.10. The illite and chlorite are distributed in sheet form, and the other minerals are distributed in granular form. The determined thresholds are then used to divide regions B and C. The content of each mineral component after division is shown in Table 3.5. According to the table, the content of each mineral component in region A, B, and C after division is basically the same. The region B and C models after 3D reconstruction are shown in Fig. 3.10. It can be found that the distribution and shape of minerals in the three regions are the same. The above results indicate that the threshold values determined are accurate.

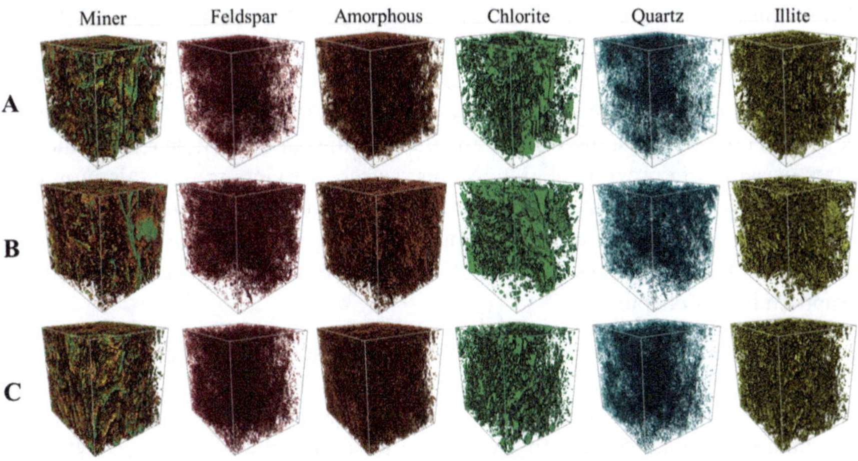


**Fig. 3.10** Three-dimensional reconstruction map of each mineral

## 3.5 Discussion and Conclusion

Pores, fractures, and mineral compositions in coal have a significant impact on the mechanical properties, seepage characteristics, and water-rock interaction of coal. Compared to the pore and mineral reconstruction methods in Table 3.1, this chapter proposes a method to determine the threshold value of various mineral components and pores inside the coal samples through CT scanning, XRD test, and NMR test. Accordingly, a three-dimensional reconstruction model of pores and mineral composition was provided. This chapter provides for the first time a classification method for different minerals in coal samples. This is crucial for revealing the mechanism of water-rock interaction in coal samples (Zhang et al., 2023c). Considering the influence of anisotropy and size effects of coal on the threshold of CT scanning reconstruction (Zhao et al., 2019b; Liu et al., 2020; Song et al., 2018), we determine the minimum and maximum representative elementary volume for reconstruction. This has also been rarely involved in previous CT reconstruction (Godinho et al., 2023; Zhou et al., 2018e; Hou et al., 2021). Besides, we selected three different regions for comparison, further verifying the reliability of our method.

However, the method proposed in this article still has some problems. The first one is that XRD analysis is destructive testing (Boral et al., 2021). Although the XRD test sample is cut to the CT scanned coal sample, the test results may still differ due to the anisotropy of the coal sample. Therefore, this article uses XRD to test the average values of coal samples from three different positions to determine the final mineral composition distribution. The second issue is the difference in accuracy between NMR and CT scans. The accuracy of CT scanning is influenced by the size of the sample, which in turn results in lower accuracy than NMR. This results in some small size pores not being recognized during the CT scanning process.

Thus, considering the issue of matching the accuracy of CT scanning and NMR, this chapter only calculates the porosity within the accuracy range of CT scanning (pore diameter greater than 0.1 μm). The final issue is to determine the density of mineral components in coal samples. Due to the difficulty in directly measuring the mineral density in coal samples, the various mineral densities in this article refer to the densities measured in previous literature; this causes overlapping intervals in mineral density. Thus, this chapter can only use the average density to distinguish minerals. In the future work, determining a simple and efficient direct measurement method for mineral density in coal sample is helpful for the further application of the method proposed in this chapter.

1. This chapter proposes a technique through CT scanning, XRD test, and NMR test. The representative elementary volume was determined through pores and minerals reconstruction of coal with different sizes. The edge lengths of the minimum representative elementary volume for pores and minerals reconstruction were 140 pixels (5 mm). While the edge lengths of the maximum representative elementary volume for pores and minerals reconstruction were 400 pixels (13.6 mm) and 580 pixels (20.7 mm), respectively. The thresholds for final determination of minerals and pores were 24,553 and 19,813 with the 300-pixel cube. The mineral content and porosity obtained using the above threshold values in three different positions are basically consistent with the measured results.
2. The pores in coal samples can be divided into connected pores and isolated pores. The number of connected pores is small, but the total volume proportion is much larger than that of isolated pores. The equivalent diameter and shape factor of a connected pores are much larger than any isolated pores. It shows that the penetration ability of the connected pores is stronger, and the channels are more tortuous and complex. Flatness can reflect the regularity of pore shape. The shape of the connected pores is close to a circle or a square. The shapes of isolated pores vary greatly, but more are close to needles.
3. X-rays have different penetrability to substances of different densities, and the greater the density, the worse the penetrability. The threshold value of various mineral components can be calculated according to the average density of the minerals and the mineral content measured by XRD. Among them, the threshold values of amorphous, feldspar, quartz, illite, and chlorite are 24,533, 25,740, 25,927, 26,826, and 28,748, respectively. The three-dimensional characterization results show that illite and chlorite are distributed in flakes, and the rest of the minerals are distributed in granular form.

BY NC ND

# Chapter 4
# Mechanical Characteristics and Damage Mechanism of Sandstone Under Long-Term Immersion

## 4.1 Introduction

The hydrogeological structure of coal mining areas in China is complex, and a large amount of mine water is bound to be produced in coal mining. At present, 1.87 $m^3$ of mine water is produced per ton of coal mined in China, and 6.88 billion $m^3$ of mine water is produced annually. According to the prediction of coal production development trend, coal mine water in China will be stable above 6 billion $m^3$ per year by 2035 (Gu et al., 2021). In the mining process, on the one hand, the formation of roof and floor rock failure zone changes the hydraulic channel between aquifers and rock strata in the mining area (Fig. 4.1a), resulting in the influx of mine water into the longwall face (Han et al., 2022b; Luo, 2020). Mine water not only weakens the coal and rock, but also corrodes the bolts and cables of roadway support to a certain extent, which weakens the support strength and makes the roadway instability (Showkati et al., 2021). On the other hand, with the application of water storage technology of mine underground reservoirs (Zhang et al., 2021c), main bearing structures (coal pillar dams, fractured rock masses, etc.) in goaf (Fig. 4.1b) will be subjected to long-term immersion effect. To guarantee the safety of longwall face mining and the stable operation of the underground reservoir, it is very important to understand the effect of long-term immersion on the physical and mechanical characteristics of rock.

In recent years, the effect of water on the physical and mechanical properties of rocks has been widely investigated in indoor experiments, such as P-wave velocity, density, tensile strength, needle penetration index, uniaxial (triaxial) compressive strength, shear strength, cohesion, and internal friction angle (Ai et al., 2021; Rabat et al., 2020; Fujii et al., 2020; Hashiba & Fukui, 2015; Chen et al., 2022a; Yao et al., 2020a, 2020b, 2020c; Liu et al., 2021; Tang, 2018). These studies found that mechanical parameters of rocks are reduced to varying degrees under the effect of water, and their physical parameters are also altered.



C. Zhang, *Water Rock Interaction in Underground Coal Mining*,
https://doi.org/10.1007/978-981-95-9957-8_4

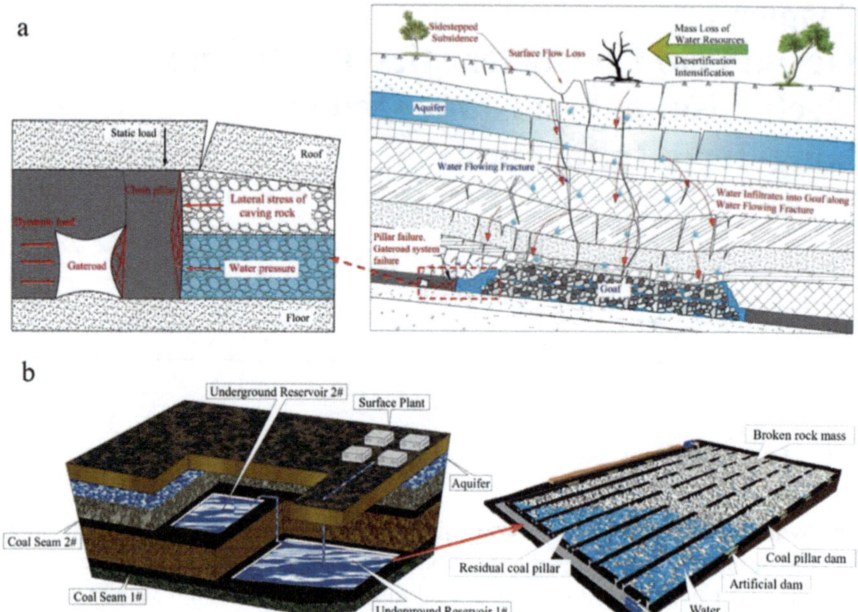


**Fig. 4.1** Long-term immersion. (**a**) Damage characteristics of overburden strata and coal pillars, (**b**) Schematic views of underground reservoir and bearing structure

The above studies have concentrated on the relationship between physical and mechanical parameters and water content states, such as natural, dry, saturated, or different water contents, while the water content remains stable after rocks reach the saturation state, ignoring the influence of immersion time on mechanical behavior. Nara et al. (2017) observed a 16% reduction in the uniaxial compressive strength (UCS) of Berea sandstone during the saturation stage, and a 30% reduction in the sandstone under 10 years of long-term immersion. Al-Bazali et al. (2011) proposed a softening factor based on the division of stress thresholds and found that the softening factor increased with increasing immersion time. In addition, variations in microstructure such as mineral composition and pore structure of rocks are similarly affected by soaking time (Ai et al., 2021). Among them, the alteration of pore structures is generally considered to be the main factor affecting the mechanical and hydraulic properties of rocks, which is the focus of this study.

A rational characterization of changes in pore structure during long-term immersion is essential to explore the weakening mechanism of water-rock interaction. Therefore, during the study of weakening mechanisms with long-term immersion, besides the characterization of mechanical strength and macroscopic fractures, the evolution of rock microstructure needs to be analyzed using a variety of techniques, such as mercury intrusion porosimetry (MIP) (Yang & Yu, 2022; Wang & Yu, 2016), scanning electron microscope (SEM) (Dehestani et al., 2020), optical microscope (OM) (Meng et al., 2021), nuclear magnetic resonance (NMR), computed tomography (CT) (Ramandi et al., 2016; Zhao et al., 2019c; Lu et al., 2021; Jiang et al.,

2014), and AE system (Gao et al., 2022; Chen et al., 2022b). SEM and OM enable direct observation of the surface morphology, composition, and fracture analysis of a sample, while MIP provides a quantitative representation of the porosity of a sample. Among them, SEM and OM enable direct observation of surface morphology, composition, and fracture analysis of samples, while MIP provides a quantitative representation of the porosity of samples. However, SEM samples require spray-gold treatment prior to testing and MIP samples require mercury intrusion during testing, which can lead to sample damage. Since CT and NMR are non-destructive techniques that can quantitatively characterize the microstructure of rock samples, they are widely applied in the evolution of pores and fractures in rocks. However, the pore size of tight sandstones is dominated by mesopores and micropores (<1 μm), and the accuracy of industrial CT cannot fully identify the pore structure of rocks. Therefore, this chapter adopts NMR techniques to quantify the evolution of sandstone pore structure during long-term immersion.

In summary, researchers are currently focusing on the mechanical properties of rocks during long-term immersion, while studies on the change in microstructure of rocks and ion concentration of soaking solution during long-term immersion are limited. In addition, the duration of long-term immersion is insufficient, with most studies concentrated on less than 1 month. Therefore, on the basis of a series of experimental tests (XRD, SEM, NMR, and hydrochemistry tests), the microscopic pore structure of sandstones is continuously characterized in this study under various soaking times. Meanwhile, the evolution of AE signals of sandstones during uniaxial compression is monitored, and the influence of water on pore structures, mechanical properties, and failure mode of sandstones is studied. Finally, the weakening mechanism of water-rock interaction during long-term immersion is explored.

## 4.2 Experimental Method

### *4.2.1 Preparation of Rock Samples*

The rock samples in this study are taken from roof strata of the Beixinyao Mine, Datong, Shanxi Province, and belong to the Permian Taiyuan Formation. In order to understand the physical and mechanical properties of sandstones during long-term immersion, the whole-rock mineral and clay mineral contents of samples are measured by X-ray diffraction. The whole-rock mineral composition is 65.2% quartz, 21.1% potassium feldspar, 3.2% calcite, 2.8% dolomite, and clay minerals including 2.3% montmorillonite, 2.1% illite, and 3.3% kaolinite (Fig. 4.2a). OM observations indicate that grain components of sandstones are dominated by quartz and potassium feldspar. The average particle size is 0.1–0.35 mm, and fillings between particles are mainly clay minerals. According to the standards set by the International Society of Rock Mechanics, sandstones are drilled along the bedding direction and machined as 50 × 100 mm (diameter × height) cylinders with a surface error within 0.3 mm and surface integrity.

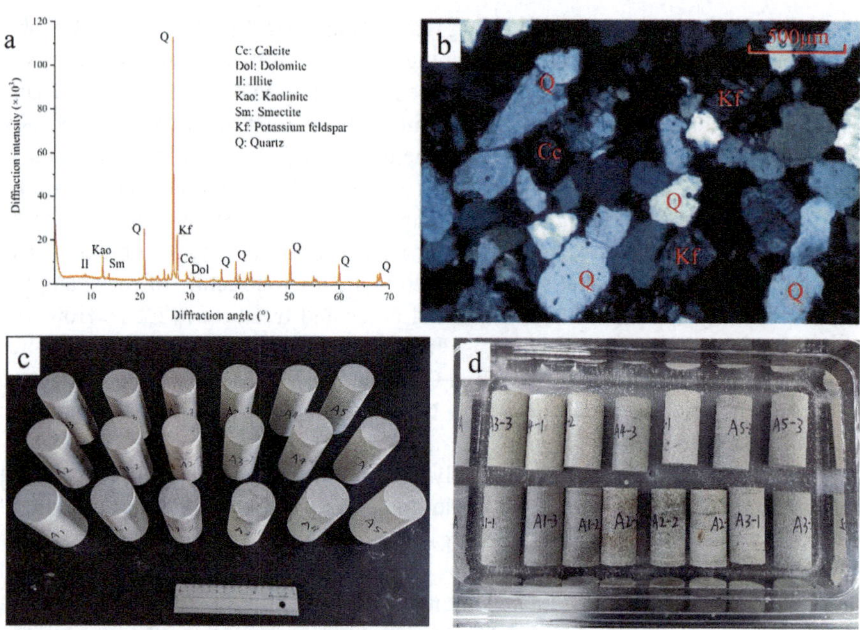


**Fig. 4.2** Mineral composition and prepared samples. (**a**) XRD spectrogram, (**b**) OM imaging, (**c**) sample photos, and (**d**) samples immersed in water

After sandstones are dried, samples are tested for P-wave velocity with a ZBL-U510 non-metallic ultrasonic detector. Twenty-one samples with similar wave velocities are selected for physical-mechanical testing to ensure the reliability of test results. All sandstones are dried at a constant temperature of 105 °C (the weight of sandstones is recorded until their weight remains constant). After drying, sandstones are sealed using preservative film. Three dry samples are immersed in water to obtain the variation curve of water content with soaking time. During the soaking, samples are weighed every 5 min. A sample is considered to be fully saturated when its weight remains stable. As shown in Fig. 4.3, the growth rate of water content of samples decreases with increasing immersion time. The reason could be that as the soaking time increases, the clay mineral particles within samples swell, resulting in a narrowing of the pore diameter, thus reducing the rate of water absorption. After 3 d of immersion, samples are saturated and the water content stabilized between 4.75% and 4.95%. All samples are divided into 6 groups of 3 samples each for different immersion times (0d, 3d, 30d, 90d, 180d, and 360d) according to the conditions of long-term immersion, as displayed in Table 4.2.

The water content of a sample is calculated as follows:

$$\omega_{\mathrm{t}} = \frac{(m_{\mathrm{t}} - m_{\mathrm{d}})100\%}{m_{\mathrm{d}}} \tag{4.1}$$

where $\omega_{\mathrm{t}}$ is the water content of samples, $m_{\mathrm{t}}$ is the weight of immersed sample, and $m_{\mathrm{d}}$ is the weight of a dry sample.

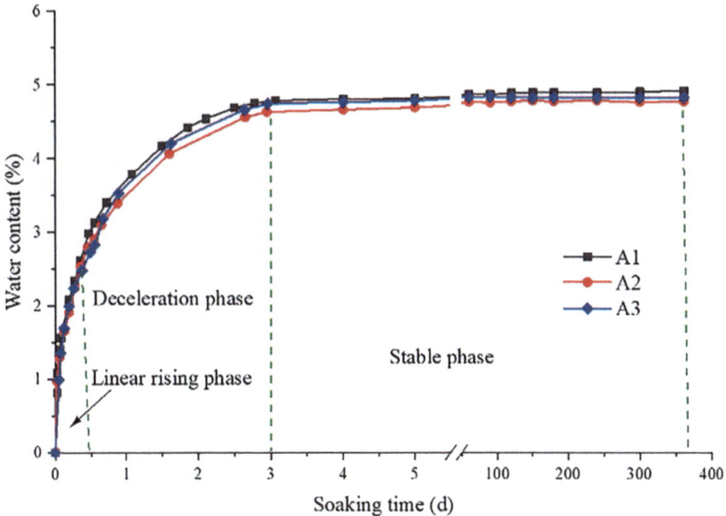


**Fig. 4.3** Moisture content of samples as a function of soaking time

### *4.2.2 Testing Equipment and Scheme*

The experimental procedure is illustrated in Fig. 4.4. Test equipment includes NMR tests, P-wave velocity tests, stress loading systems, and AE systems. To quantitatively describe the pore characterization of samples at different soaking times, NMR and P-wave velocity tests are performed on three samples from group A6 after immersion for set times. The relaxation time T2 of hydrogen-containing fluid in sample pores is measured by a MesoMR23-060H-I low-field NMR system. Among them, dry samples are placed in a vacuum and pressure saturation apparatus to saturate samples, and then NMR tests are carried out, while saturated samples can be used directly for NMR testing. Before tests, residual water is removed from the sample surface to reduce interference with the NMR results. When NMR tests are completed, samples continue to soak until all designed immersion times are tested.

During the soaking, we perform a hydrochemistry analysis of soaking solution. The pH value and total dissolved solids (TDS) in the immersion solution are measured by a high-precision pH gauge and a conductivity gauge with an accuracy of 0.01 and 0.1 mg/L, respectively. The ionic concentration is determined by ion chromatography with an accuracy of 0.01 mg/L. The $SiO_2$ content is determined by a silicon molybdenum yellow photometric method with an accuracy of 0.1 mg/L. According to a set soaking time of samples, the process is carried out five times. The volumes of sandstones and distilled water are selected in a ratio of 1:2 in order to avoid the influence of factors other than soaking time on the hydrochemistry analysis. Among them, the volume of each group of sandstone is approximately 600 $cm^3$, and the soaking solution is taken as 1200 mL at a time for hydrochemistry analysis.

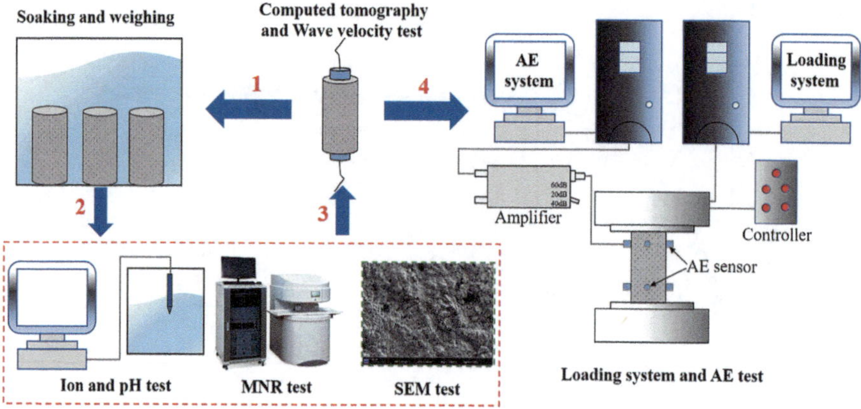


**Fig. 4.4** Diagram of experimental procedure

After the immersion is completed, the MTS C45.104 universal testing machine is used in this experiment to load samples with different immersion times. This loading system has a maximum load capacity of 300 kN and a measurement accuracy error of less than 0.5%. The AE device is a PCI-Express 8 AE system with a NANO-30 sensor. Before starting the tests, the loading rate of samples is set to 0.06 mm/min, and the six sensors are installed on the surface of samples using a coupling agent, as displayed in Fig. 4.4. The preamplifier and acquisition threshold for AE signals are both set to 40 dB, and the acquisition frequency is 1 MHz. AE system and universal testing machine start recording data simultaneously until the sample is destroyed.

## 4.3 Physical and Mechanical Properties of Sandstone with Different Soaking Time

### *4.3.1 Analysis of P-Wave Velocity*

Figure 4.5 shows the time dependence of P-wave velocity of sandstones. As shown in Fig. 4.5, the P-wave velocity shows a variation pattern of increasing and then decreasing with increasing immersion time. When a sample is in an unsaturated state, the saturation affects the variation of P-wave velocity (Wang et al., 2020c). As can be seen from Fig. 4.5, the growth rates of wave velocity and water absorption of samples show the same trend. Since the propagation velocity of waves in water is greater than that in air, for sandstones with high water content, the gradual filling of pores by water leads to an increase in P-wave velocity (Ai et al., 2021). When a sample reaches saturation state, P-wave velocity and water content of

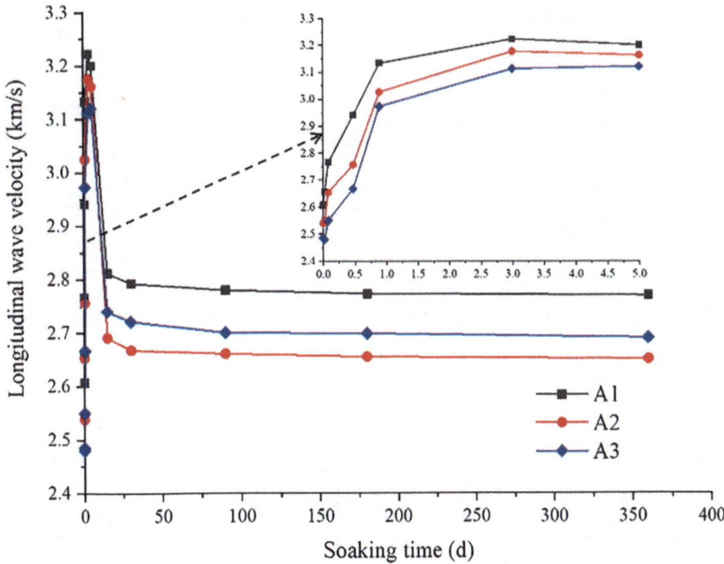


**Fig. 4.5** Curve of P-wave velocity with soaking time

samples reach a maximum. As the immersion time continues to increase, the P-wave velocity tends to decrease. The reason for this may be, on the one hand, the adsorption swelling effect of clay minerals (montmorillonite, illite). The entry of water molecules between the montmorillonite crystal layers increases the spacing of mineral particles, which leads to a decrease in wave velocity (Zhu et al., 2022). On the other hand, long-term immersion leads to the dissolution of mineral components and the shedding of rock debris, which alters the pore structure characteristics of samples and ultimately causes a decrease in P-wave velocity (Wang et al., 2015).

### *4.3.2 Microstructural Descriptions*

1. Surface morphology distribution
   The change in rock microstructure determines the mode of macroscopic failure. To further understand the change in microstructure of sandstones during long-term immersion, SEM tests are carried out on sandstones with different soaking times, as shown in Fig. 4.6. Cracks are not produced throughout the immersion process, and the pore characteristics are always predominantly micropore. The dry sandstone has a flat, continuous surface and a tight matrix in which mineral grains are cemented by clay minerals (Fig. 4.6). When samples are soaked for 3 d, the sample surface transforms from flat to rough and the number of macropores in samples begins to increase. In addition, due to the swelling properties of

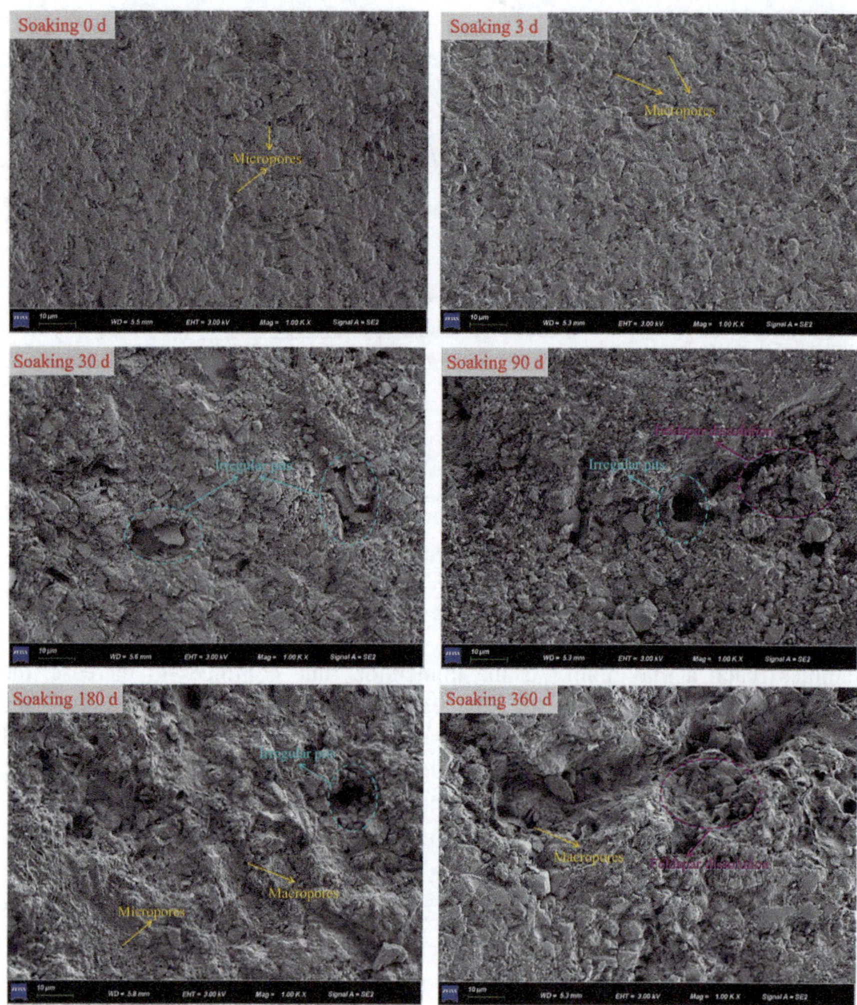


**Fig. 4.6** SEM imaging of sandstones at different soaking times

clay minerals, the number of micropores is significantly less than that of dry sandstone. In the SEM image at a soaking time of 30 d, the boundary of mineral grains is visible and rock debris is partially exfoliated to form irregular pits. As the soaking time continues to increase to 90 d, the potassium feldspar undergoes corrosion, forming large corrosion pores and leaving feldspar remains (Ma et al., 2017; Jiang et al., 2022b; Sun et al., 2022). This phenomenon is further aggravated during subsequent soaking. When the soaking time reaches 360 d, pore sizes become larger and interconnected, the number of large pores grows, and the surface becomes uneven.

2. Pore structure measurement
The T2 (transverse relaxation time) distribution curve of NMR provides information on the pore distribution of rocks and the characteristic parameters of the pore structure. Based on the theory of NMR testing of rocks, it is known that the T2 value of rocks is proportional to the pore throat radius. Among them, the longer the T2 relaxation time, the larger the pore size of a sample, and the higher the T2 spectral peak, the more the number of pores (Meng et al., 2022; Li et al., 2021a). As can be seen from Fig. 4.7, the T2 spectra of samples at different soaking times show a bimodal distribution, indicating that various pore sizes are developed within samples. Due to the porous media properties of sandstones, the pore size of samples can be divided into three grades based on the principle of NMR measurement of pores (Wang et al., 2022) (Table 4.1).

The T2 spectral area quantitatively characterizes the variation of pore size and number within rocks (Zhao et al., 2017a). Figure 4.8 shows the distribution of pore proportion and total pore signal of samples with the increase of soaking time. The pore size distribution of samples at different soaking times is consistent. Among them, the number of micropores accounts for 83.3–90.5% of the total number of pores, the number of mesopores accounts for 6.6–10.8%, and the number of macropores accounts for only 2.9–5.9%. As the soaking time increases, the pore distribution proportion of micropores tends to decrease, while the pore

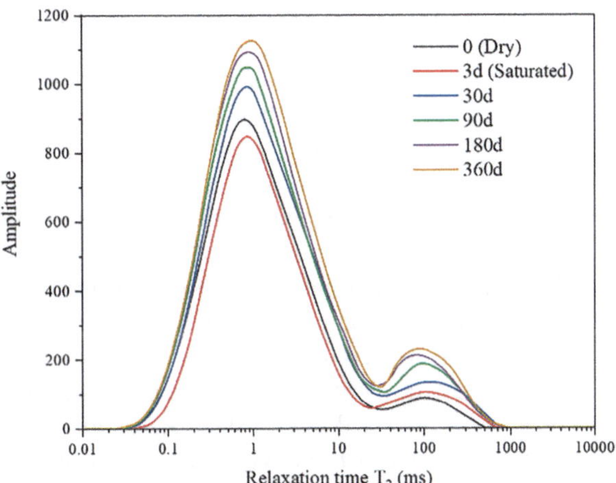


**Fig. 4.7** Variation of transverse relaxation T2 in sandstones at different soaking times

**Table 4.1** T2 spectra corresponding to pore size grades

| Pore size grade | Range of pore sizes | T2 value |
|---|---|---|
| Micropore | <0.1 μm | <100 ms |
| Mesopore | 0.1–1 μm | 10–100 ms |
| Macropore | >1 μm | >100 ms |

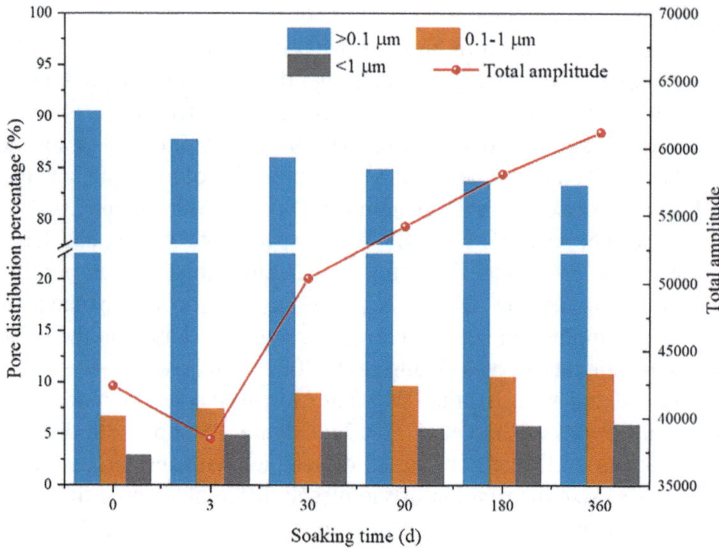


**Fig. 4.8** Pore structure characteristics of sandstones at different soaking times

distribution proportion of mesopores and macropores gradually grows. Compared with the pore distribution proportion of dry sandstones, the pore distribution proportion of micropores decreases by 7.2% after 360 d of immersion, while the pore distribution proportion of mesopores and macropores increases by 4.2% and 3.0%, respectively. However, the total pore amplitude of samples shows a trend of decreasing and then increasing with the increase of soaking time (Fig. 4.8). The total pore amplitude of samples decreases from 42321.2 to 38407.3 when the immersion time is increased from 0 to 3 d (saturated), a decrease of 9.2% compared to dry samples. The total pore amplitude of samples soaked for 360 d increases by 44.4% as the immersion time continues to increase. The results show that immersion for 3 d (saturated state) is a turning point. The total pore signal of samples reduces in the unsaturated stage, in which the adsorption swelling effect of clay minerals mainly occurs. After a sample reaches saturation state, the porosity of sandstones gradually increases as the cemented minerals begin to be dissolved and rock debris is shed. The above process of pore size variation is consistent with the observation of SEM images.

### 4.3.3 Mechanical Properties and AE Responses

Figure 4.9 shows the stress-strain curves for sandstones at different soaking times. As can be observed from Fig. 4.9a, there is diversity in stress-strain curves of sandstones at different soaking times. The slope of stress-strain curves decreases with the increase of soaking time before the peak stress, indicating that the deformation modulus of sandstones gradually decreases during long-term immersion. The post-peak curve is presented in three main forms, namely linear, stepped, and smooth. Among them, the dry sandstone is typical of linear damage. When the stress reaches peak strength, the stress-strain curve decreases rapidly and the sandstone shows brittle characteristics. With the increase of soaking time, the stress-strain curve gradually transforms from a stepped shape to a smooth shape in the post-peak stage, and the post-peak gradient tends to decrease (Tang et al., 2021a). When the soaking time increases from 0 to 360 d, the average peak stress in sandstones decreases from 106.4 to 18.1 MPa, a reduction of 83.0% compared to the peak stress in dry sandstones. The drop rate of peak stress gradually decreased during long-term immersion (Fig. 4.9c). The average peak strains of sandstones at different water immersion times are 0.0172, 0.0128, 0.0116, 0.0105, 0.0112, and 0.0119, respectively, and it can be found that peak strains first fall and then rise with the increase of soaking time (Fig. 4.9d and Table 4.2).

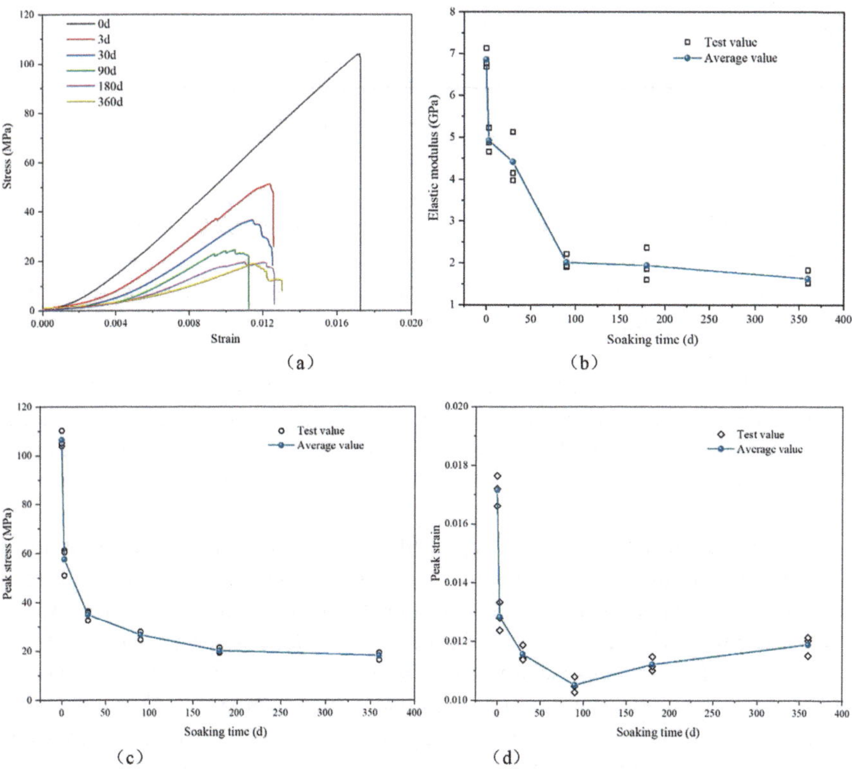


**Fig. 4.9** Variation of stress-strain curves and mechanical parameters of samples at different soaking times. **(a)** Stress-strain curve, **(b)** Elastic modulus, **(c)** Peak stress, and **(d)** Peak strain

**Table 4.2** Numbering and mechanical parameters of sandstones at different immersion times

| | | Peak stress (MPa) | | Peak strain | | Elastic modulus (GPa) | |
|---|---|---|---|---|---|---|---|
| Sample number | Soaking time (d) | Tested | Average | Tested | Average | Tested | Average |
| A1 | 0 | 103.9 | 106.4 | 0.01569 | 0.01665 | 6.69 | 6.87 |
| A2 | | 110.2 | | 0.01765 | | 6.78 | |
| A3 | | 105.1 | | 0.01661 | | 7.13 | |
| A1–1 | 3 | 51.8 | 57.8 | 0.01238 | 0.01284 | 5.23 | 4.93 |
| A1–2 | | 61.2 | | 0.01334 | | 4.89 | |
| A1–3 | | 60.5 | | 0.0128 | | 4.66 | |
| A2–1 | 30 | 36.4 | 34.9 | 0.01144 | 0.01157 | 4.15 | 4.42 |
| A2–2 | | 35.8 | | 0.01138 | | 5.13 | |
| A2–3 | | 32.6 | | 0.01188 | | 3.98 | |
| A3–1 | 90 | 24.6 | 26.6 | 0.01048 | 0.01052 | 1.9 | 2.02 |
| A3–2 | | 27.2 | | 0.0108 | | 2.22 | |
| A3–3 | | 27.9 | | 0.01027 | | 1.93 | |
| A4–1 | 180 | 19.3 | 20.1 | 0.01101 | 0.01122 | 2.37 | 1.94 |
| A4–2 | | 19.6 | | 0.01116 | | 1.86 | |
| A4–3 | | 21.4 | | 0.01148 | | 1.6 | |
| A5–1 | 360 | 18.0 | 17.9 | 0.01152 | 0.0119 | 1.83 | 1.63 |
| A5–2 | | 19.3 | | 0.01203 | | 1.52 | |
| A5–3 | | 16.4 | | 0.01214 | | 1.53 | |

During uniaxial compression, samples usually undergo five phases, i.e., compression phase (A–B), elastic phase (B–C), stable fracture development phase (C–D), unstable fracture extension phase (D–E), and post-peak failure phase (E–F). Understanding the evolution of rocks in various phases during long-term immersion plays a vital role in preventing rock failure. The stress corresponding to a specific time (or strain) can be obtained through the stress-strain curve, but it is impossible to accurately divide each phase in the process of rock loading. Since AE signals generated by various loading phases are different, the combination of AE technology can provide a basis for the division of phases (Tang et al., 2022).

The fracture process of a rock is a result of continuous accumulation, transformation, and intermittent release of internal strain energy. The AE count reflects the degree of internal damage in sandstones during loading (Zhao et al., 2016; Ohtsu, 1991). Figure 4.10 illustrates time sequence variations of stress, AE counts and cumulative AE counts of sandstones at different immersion times. During the compression phase (A–B), dry sandstones produce a small amount of AE counts, while sandstones with longer soaking time generate fewer AE counts. Dry sandstones generate a small amount of AE signals due to the occlusion and friction of pre-existing pore structures within sandstones in the process of compression and closure. For sandstones with longer immersion times, water can reduce occlusion and

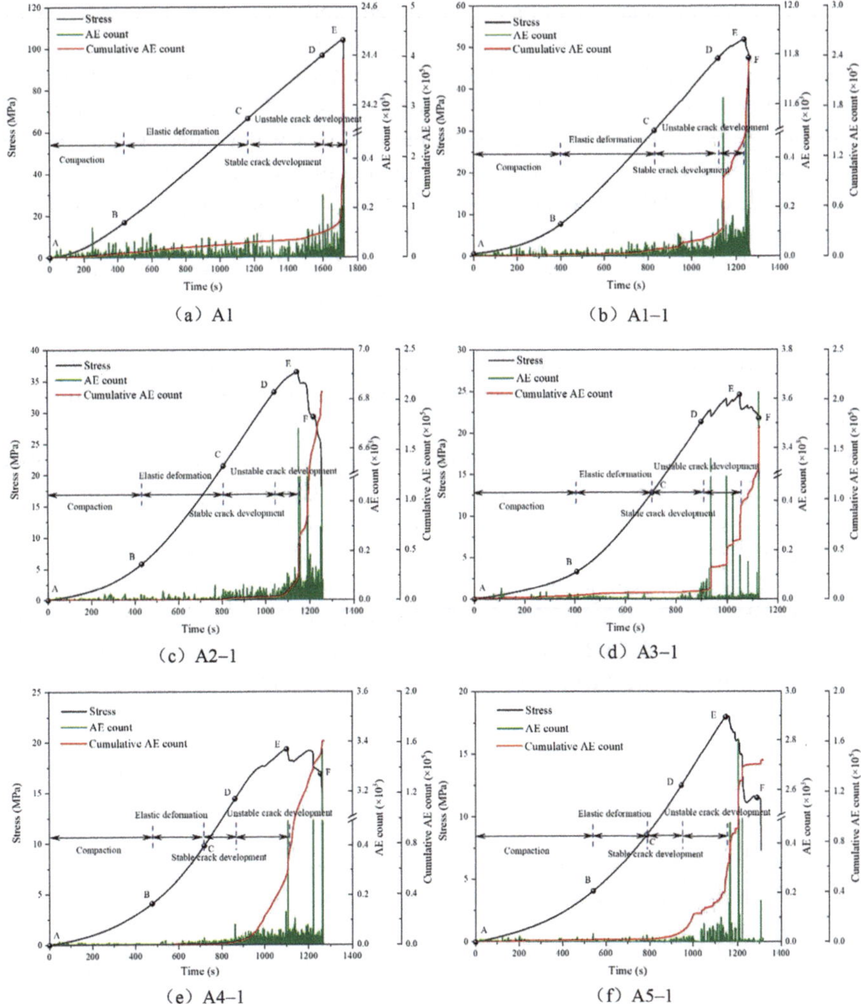


**Fig. 4.10** Time sequence variations of stress, AE counts, and cumulative AE counts. (**a**) A1, (**b**) A1–1, (**c**) A2–1, (**d**) A3–1, (**e**) A4–1, and (**f**) A5–1

friction of pre-existing pore structures within sandstones, and the presence of water may cause AE signals to decay in the transfer process, resulting in a reduction of AE signals. As the loading progresses, especially in stable fracture development phase (C–D) and unstable fracture extension phase (D–E), AE counts and cumulative AE counts increase significantly at different soaking times, indicating that the expansion and penetration of fractures within sandstones promote the generation of AE signals. However, the peak AE counts and cumulative AE counts decrease with increasing immersion time. The peak AE count of dry sandstone is

24,268, while the other five groups of saturated sandstones with different soaking times are 11,624, 6678, 3625, 3366, and 2804, a reduction of 52.1%, 72.5%, 85.1%, 86.1%, and 88.4%, respectively, compared to dry sandstone. A cumulative AE count of sandstones at 360 d immersion time is 12,812, a reduction of 96.8% compared to dry sandstones. The results indicate that water entering the pores of sandstones weakens the cementing effect of clay minerals between particles, causing a reduction in UCS of samples. Meanwhile, the strain energy stored in elastic phase of long-term immersed sandstone is less than that of dry sandstone, and the presence of water absorbs the elastic energy released by fracture propagation, which eventually leads to a decrease in AE signals of sandstones during loading (Li et al., 2021b).

To better compare the distribution of each phase of sandstones at different immersion times, we adopt a normalization treatment for strain data at each stage, where the vertical coordinate is the ratio of stage strain to peak strain, as shown in Fig. 4.11. For sandstones with long-term soaking, the increase in size and number of pores and the weakening of cementing minerals result in a greater strain being required to close the pore structure during a compression phase. However, the stage ratio of sandstones in elastic phase and stable fracture development phase decreases with increasing immersion time. For the sandstone soaked for 360 d, the stage ratios in elastic phase and stable fracture development phase are 0.216 and 0.131, respectively, a reduction of 53.2% and 51.6% compared to dry sandstone. The results indicate that the elastic energy stored in a sandstone during loading decreases with

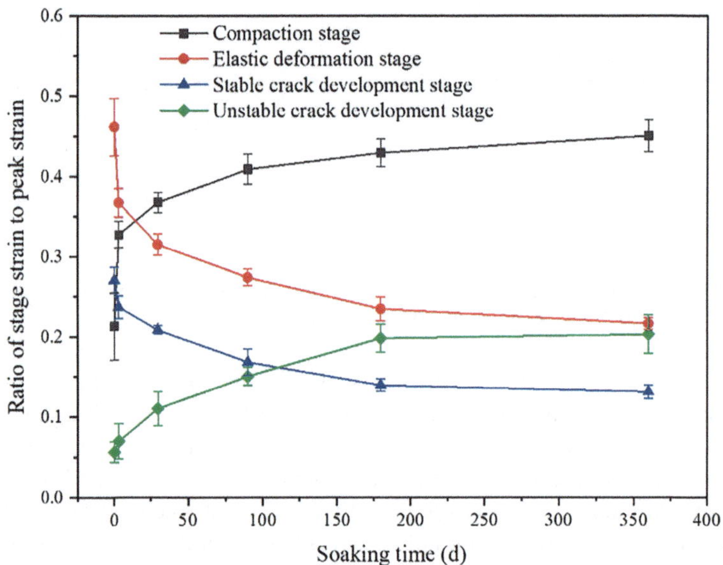


**Fig. 4.11** Variation of the ratio of stage strain to peak strain with soaking time

increasing immersion time. Compared with dry sandstone, the stage ratio in unstable fracture extension phase increases by 0.014, 0.054, 0.095, 0.142, and 0.147 under different soaking times, indicating that long-term water-rock interaction makes sandstones enter the plastic stage earlier and the plastic characteristics become more obvious with increasing soaking time.

A rock sample is accompanied by the creation and expansion of fractures in the process of loading, and the identification of fracture types can effectively predict the mode of final instability of rock mass. Shear cracks produce low average frequency (AF) and long rise times of acoustic emission signals, while the opposite is true for tensile cracks (Farhidzadeh et al., 2014; Aggelis et al., 2012; Yao et al., 2019a, 2019b). Therefore, the ratio of RA to AF is used to indicate the mode of fracture (Chen et al., 2022a), as shown in Fig. 4.12. Among them, the expressions of RA and AF are as follows:

$$RA = \frac{\text{Rise time}}{\text{Amplitude}} \tag{4.2}$$

$$AF = \frac{\text{Count}}{\text{Duration}} \tag{4.3}$$

Figure 4.12 shows the classification of AE signals and the failure mode of a sandstone at different soaking times. According to the phase division of the stress-strain curve, five different colors are used to record the AE data of each phase. Among them, the maximum values for RA and AF are 2000 ms/v and 1000 kHz, respectively. As can be noted from Fig. 4.12, long-term immersion has a significant effect on the failure process of sandstones. For dry sandstone, the damage ratio starts to increase steeply at 1560s, showing brittle fracture characteristics. Tensile fractures are 2.3 times larger than shear fractures after loading. The macroscopic failure mode of A1 sandstone mainly includes two tensile crack planes and one shear crack plane, which is consistent with RA/AF damage results (Fig. 4.12a). When the soaking time is 3 d, the shear fractures are always smaller than the tensile cracks during loading, and fractures develop in a similar way to dry samples, but sandstones begin to show post-peak characteristics at this time (Fig. 4.12b). When soaking time reaches 30 d, the damage ratio of tensile and shear fractures are only 0.18% and 0.13% before the elastic stage, while shear fractures are larger than tensile fractures after the elastic phase, and eventually two types of fractures are about equal in number (Fig. 4.12c). During immersion from 90 to 360 d, microfractures in sandstones are dominated by shear fractures, which increase by 18.0%, 19.6%, and 32.6%, respectively, compared to the damage ratio of tensile fractures, indicating that the damage ratio of shear fractures continued to rise with increasing immersion time. In addition, with the increase of soaking time, the macroscopic failure mode of a sandstone changes from a combination of shear and tensile crack surfaces to a single shear crack surface, as shown in Fig. 4.12d–f.

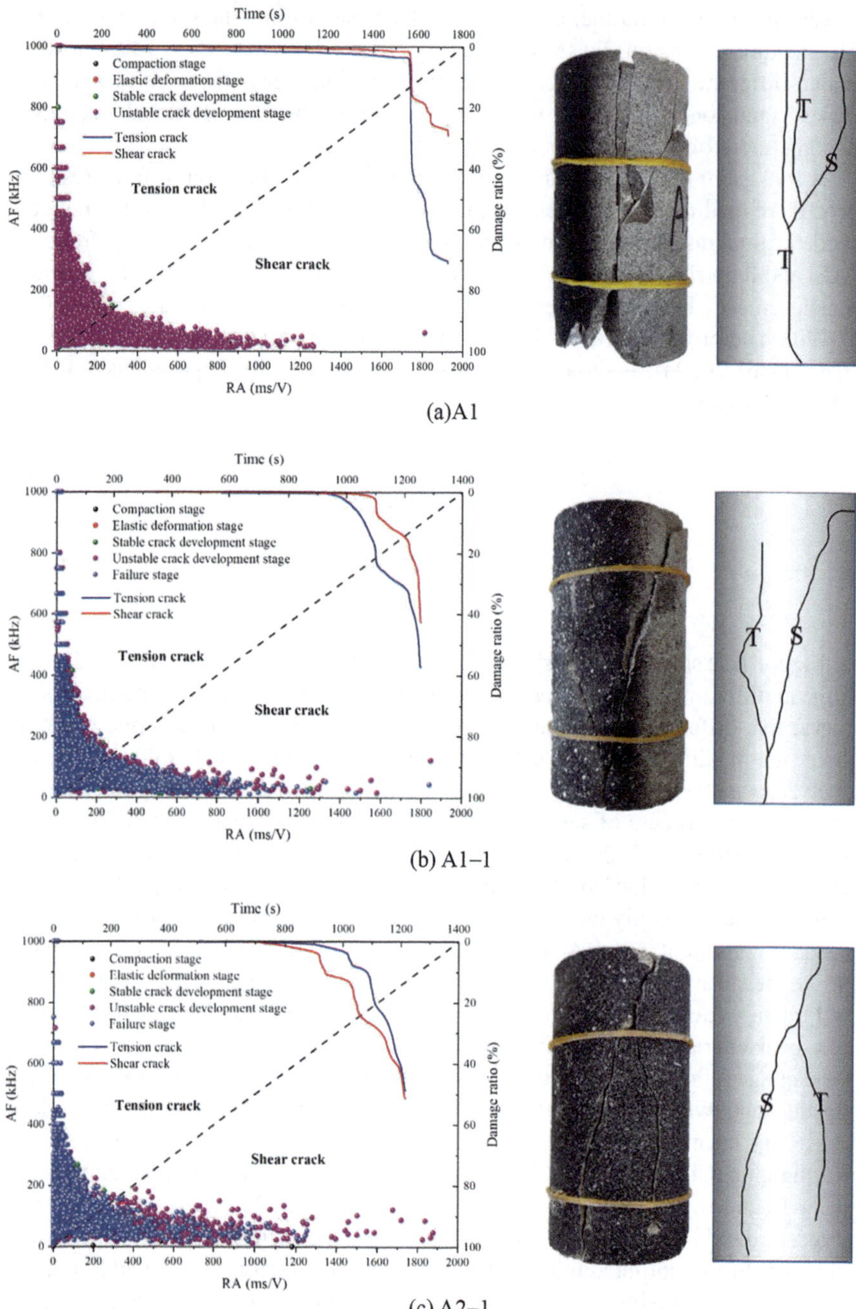


**Fig. 4.12** Variation of RA/AF and failure modes in sandstone at different immersion times. (**a**) A1, (**b**) A1–1, (**c**) A2–1, (**d**) A3–1, (**e**) A4–1, (**f**) A5–1

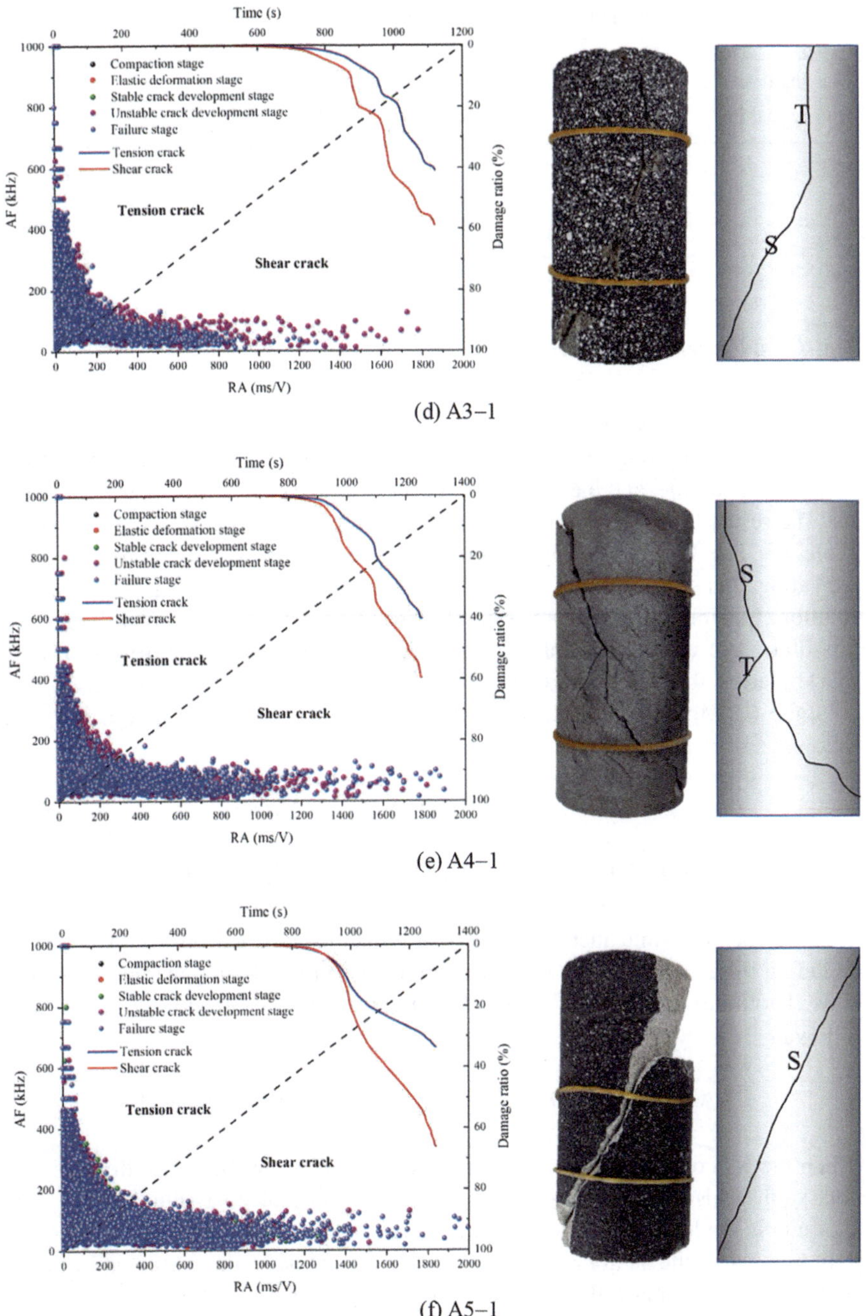


**Fig. 4.12** (continued)

**Table 4.3** Variation in ion concentration during long-term immersion

| Soaking time (d) | pH value | TDS (mg/L) | Ion mass concentration (mg/L) | | | |
|---|---|---|---|---|---|---|
| | | | $Mg^{2+}$ | $K^+$ | $Ca^{2+}$ | $SiO_2$ |
| Before soaking | 7.04 | 10.3 | 0.21 | 0.72 | 0.2 | 0 |
| 3 | 7.75 | 45.3 | 0.86 | 1.09 | 4.9 | 0.4 |
| 30 | 8.01 | 86 | 1.37 | 2.99 | 9.1 | 3.2 |
| 90 | 7.95 | 110.9 | 1.97 | 2.94 | 12.6 | 3.6 |
| 180 | 7.88 | 133 | 2.38 | 2.85 | 16.3 | 3.8 |
| 360 | 7.88 | 145.4 | 2.43 | 2.9 | 18 | 4 |

### *4.3.4 Mineral Dissolution*

To reveal the influence of physicochemical reactions on mechanical properties of sandstones during long-term immersion, distilled water is used to immerse for a predetermined time, and the soaking solution is subjected to hydrochemistry analysis. The results of hydrochemistry tests at different immersion times are shown in Table 4.3. The TDS of soaking solution increases with increasing immersion time, indicating that minerals in sandstone are continuously dissolving in immersion solution. Eslami et al. (2010) classified dissolution reactions into total dissolution (dissolution of calcite and dolomite) and non-total dissolution (corrosion of feldspar) based on the type of products. The dissolution reaction of calcite ($CaCO_3$) and dolomite ($CaMg(CO_3)_2$) produces $Ca^{2+}$ and $Mg^{2+}$, and the reaction equation is as follows:

$$\mathrm{CaCO_3} + CO_2 + \mathrm{H_2O} \rightarrow Ca^{2+} + 2HCO_3^- \tag{4.4}$$

$$\mathrm{CaMg}\left(CO_3\right)_2 + 2CO_2 + 2\mathrm{H_2O} \rightarrow Ca^{2+} + Mg^{2+} + 4HCO_3^- \tag{4.5}$$

As immersion time increases, feldspar minerals produce corrosion products, including illite and kaolinite. Among them, potassium feldspar generates clay minerals (kaolinite) in the early stage of corrosion, and the reaction equation is as follows:

$$2\mathrm{KAlSi_3O_8} + 3\mathrm{H_2O} = Al_2Si_2\mathrm{O_5}\left(OH\right)_4 + 4SiO_2 + 2OH^- + 2\mathrm{K}^+ \tag{4.6}$$

When the $K^+$ concentration in the pore fluid reaches above the saturation point of potassium feldspar, additional $K^+$ is the main source of potassium in illite, and kaolinite is gradually converted to illite (Berger et al., 1997). When the $K^+$ concentration is below the saturation point of potassium feldspar, the corrosion of potassium feldspar generates illite, and the reaction equation is as follows:

$$3\mathrm{Al_2}Si_2\mathrm{O_5}\left(OH\right)_4 + 2\mathrm{K}^+ = 2KA\mathrm{l_3}Si_3\mathrm{O_{10}}\left(OH\right)_2 + 2\mathrm{H}^+ + 3\mathrm{H_2O} \tag{4.7}$$

$$\mathrm{KAlSi_3O_8} + Al_2Si_2\mathrm{O}_5\left(OH\right)_4 = KAl_3Si_3\mathrm{O}_{10}\left(OH\right)_2 + 2SiO_2 + \mathrm{H_2O} \quad (4.8)$$

As can be seen from Table 4.3, the pH value of soaking solution presents a trend of increasing and then decreasing with increasing immersion time. During the saturation stage of sandstones, the concentration of $Mg^{2+}$ and $Ca^{2+}$ increases by 0.75 and 4.7 mg/L, respectively, while the concentration of K+ increases by only 0.32 mg/L, indicating that the dissolution reactions of calcite and dolomite mainly occur during this stage (Eqs. 4.4 and 4.5). In addition, the $HCO_3^-$ produced by the above reaction leads to an increase in the pH value of the soaking solution. When sandstones are soaked to 30 d, $Mg^{2+}$ and $Ca^{2+}$ are still increasing steadily, while the concentration of $K^+$ reaches a maximum, increasing by 174.3% compared to the saturation stage. During this stage, the potassium feldspar starts to undergo a corrosion reaction, which produces $OH^-$, leading to a still-increasing pH value in the soaking solution. As the soaking time continues to increase, the $K^+$ concentration and pH value in the soaking solution first decrease and then stabilize at 2.9 mg/L and 7.88, respectively. The reason may be that the $K^+$ concentration in the soaking solution is higher than the saturation point of potassium feldspar during this stage, and the conversion of kaolinite to illite produces $H^+$ and consumes part of $K^+$ (Eq. 4.7). Therefore, it can also be deduced that the critical concentration of $K^+$ in this experiment is 2.9 mg/L. Throughout the soaking process, a part of generated $SiO_2$ dissolves in the soaking solution as a colloid, and the other part adheres to the surface of pore wall and precipitates in the soaking solution.

## 4.4 Long-Term Water Immersion Weakening Mechanism

The reduction in strength of sandstone due to the presence of water is common. Previous scholars conducted numerous studies on the weakening mechanisms of water-rock interaction (Cai et al., 2020a, 2020b; Li et al., 2020a), which include five main mechanisms: increased pore pressure, increased stress corrosion, decreased intergranular cohesion and friction, decreased fracture energy, and swelling and dissolution of mineral components. However, no single mechanism can fully explain the weakening effect of immersion. Wasantha and Ranjith et al. (2014) observed that immersion weakening may be the result of a combination of two or more mechanisms. The main reason is that water-rock interaction is affected by the microstructure of sandstone, e.g., pore structure characteristics, mineral composition, cementation patterns between particles, and particle size. The water weakening properties of the same type of sandstones are highly variable, as illustrated in Table 4.4.

Previous studies have shown that different types of sandstones can be divided into disintegration sandstones, swelling sandstones, and dissolution sandstones according to the type of water-rock interaction during long-term immersion (Table 4.5). As can be seen from Table 4.4, the composition of different types of

**Table 4.4** Summary of laboratory tests of different moisture contents' effect on mechanical characteristics of sandstone

| Rock type (Region) | Treatment methods | Uniaxial compressive strength (MPa) | | Weakening coefficient | Porosity (%) | Reference |
|---|---|---|---|---|---|---|
| | | Dry | Saturated | | | |
| Sandstone (Sichuan, China) | $\omega$: 0%, 1.03%, 2.04%, 3.14% | 42.35 | 23.53 | 0.556 | 6.40 | Li et al. (2021) |
| Black sandstone (Sichuan, China) | $\omega$: 0%, 0.32%, 0.34%, 0.7%, 0.9%, 1.2%, 1.62% | 124.64 | 61.25 | 0.491 | 4.70 | Tang et al. (2018) |
| Sandstone (Australia) | $\omega$: 0%, 1.61%, 2.42%, 3.81%, 4.74%, 6.88% | 47.01 | 13.61 | 0.290 | 18 | Masoumi et al. (2017) |
| Mudstone (Xinjiang, China) | $\omega$: 0%, 1.0%, 2.0%, 3.0%, 4.0%, 5.1%, 6.0% | 23.38 | 4.78 | 0.204 | 9.42 | Liu et al. (2021) |
| Sandstone (Shandong, China) | $\omega$: 0%, 0.8%, 1.6%, 2.4%, 3.2% | 90.67 | 33.74 | 0.372 | 8.05 | Sun et al. (2021a) |
| Sandstone (Shanxi, China) | $\omega$: 0%, 1.13%, 1.58%, 2.52%, 3.63% | 111.47 | 76.47 | 0.686 | 4.75 | Chen et al. (2021b) |
| Red sandstone (Hunan, China) | $\omega$: Dry, saturated (3.40%) | 108.17 | 56.57 | 0.523 | 11.60 | Tang et al. (2018) |
| Sandstone (China) | $\omega$: 0%, 0.50%, 1.50%, 2.50% | 115.90 | 52.57 | 0.454 | 5.38 | Sun et al. (2021b) |
| Sandstone (Isparta, Turkey) | $\omega$: Dry, saturated (2.37%) | 118.8 | 54.6 | 0.460 | 5.3 | Erguler and Ulusay (2009) |
| Red sandstone (China) | $\omega$: Dry, natural (0.65%), saturated (2.1%) | 97.5 | 86.3 | 0.885 | 4.71 | Luo (2020) |
| Red sandstone (Hunan, China) | $\omega$: 0%, 0.7%, 1.6%, 2.6%, 3.5%, 4.6%, 4.7% | 73.65 | 50.25 | 0.682 | 12.60 | Yu et al. (2019) |
| Red sandstone (Jiangxi, China) | $\omega$: Dry, natural (1.46%), saturated (2.76%) | 97.80 | 56.22 | 0.575 | 2.80 | Zhao et al. (2021a) |
| Sandstone (Yunnan, China) | $\omega$: 0%, 1%, 2%, 3.5% | 66.75 | 46.80 | 0.701 | 9.30 | Zhou et al. (2016) |

(continued)

**Table 4.4** (continued)

| Rock type (Region) | Treatment methods | Uniaxial compressive strength (MPa) | | Weakening coefficient | Porosity (%) | Reference |
|---|---|---|---|---|---|---|
| | | Dry | Saturated | | | |
| Sandstone (Ankara, Turkey) | $\omega$: Dry, saturated (3.7%) | 61.1 | 37.9 | 0.620 | 8.9 | Erguler and Ulusay (2009) |

**Table 4.5** Water-rock weakening effect in different types of sandstone

| Sandstone type | Mineralogical composition (%) | | | | | | Water-rock interaction type | Reference |
|---|---|---|---|---|---|---|---|---|
| | Q | Kf | Cc | Il | Kao | Sm | | |
| Datong (this study) | 65.2 | 21.1 | 3.2 | 2.1 | 3.3 | 2.3 | Corrosion | – |
| Lanzhou | 28.28 | 25.8 | 17.95 | 6.23 | 2.82 | 18.65 | Disintegration | Azhar et al. (2020) |
| Zhuzhou | 31.01 | 6.28 | 20.13 | – | 32.33 | 8.37 | Disintegration | Fan et al. (2022) |
| Ordu | 6.1 | 8.2 | 22.7 | 4.13 | 3.54 | 48.38 | Swelling | Bilir and Sarıgül (2021) |
| Yichang | 48 | 6 | 31 | – | – | 8 | Dissolution | Xie et al. (2018a) |

sandstone usually consists of feldspar, calcite, clay minerals, and quartz grains, but there are significant differences in their content. Among them, the disintegrated sandstone has a quartz content of about 30%, dissolved minerals (feldspar and calcite) of 30–40% and clay minerals of 30–40%. The mineral composition of swelling sandstone is dominated by clay minerals, with a content of 55%, while the quartz content is only 6.1%. For dissolved sandstones, the main mineral composition is quartz (50–60%), while the clay mineral content is less than 10%. Since most sandstones contain the same mineral composition, we can speculate that the mechanism of water-rock weakening may be the same for different types of sandstone. However, the difference in microstructure and the length of immersion time dictate that some mechanisms may be more important than others.

In general, the pore pressure effect is positively correlated with the pore volume and connectivity of the sandstone (Hawkins & McConnell, 1992). Since the sandstone matrix in this experiment is tight, the porosity is only 3.8–4.7%. Sandstones with different immersion times are tested for mechanical strength considering only the effect of axial stress; the pore pressure can be released laterally along a sample. Stress corrosion is prevalent in silicates, where the support elements (Si-O bonds) at the crack tip are converted to lower strength hydrogen bonds, reducing the energy required for crack expansion (Xie et al., 2018b). Brantut et al. (2014) found that stress corrosion mechanisms mainly occur at strain rates less than $10^{-7}$ s$^{-1}$. In contrast, sandstones in this experiment have a strain rate of $10^{-5}$ s$^{-1}$ during loading, which may inhibit the extent of stress corrosion. Therefore, it can be assumed that the pore pressure effect and the stress corrosion effect make a small contribution to the strength reduction of sandstone in this experiment.

For the sandstones selected for this study, there are differences between the strength and failure modes under different immersion times. Among them, the UCS of sandstones decrease by 47.4% and 83.0% at the saturation stage and at 360 d of immersion, respectively. The NMR and hydrochemistry test results (Fig. 4.8 and Table 4.3) show that the swelling of clay minerals leads to a decrease in the number of micropores during the saturation stage, while the dissolution of calcite cement results in an increase in the number of mesopores and macropores. The total pore signal of saturated sandstone drops to a minimum value, indicating that the swelling of clay minerals is dominant at this stage. Zhu et al. (2022) noted that the adsorption of water molecules on the crystal surface or between the crystal layer of clay minerals is the essence of sandstone swelling. When water is present, some of water molecules adsorb to the crystal surface, reducing the crystal surface energy and thus promoting the development of intergranular cracks. Another part of water molecules enters between crystal layers, resulting in a change of interlayer connections and an increase of interlayer spacing, which reduces cohesion and may even lead to disintegration of the rock sample in some cases (Hu et al., 2017). The quartz grains in sandstones are mainly cemented by potassium feldspar (25.1%) and clay minerals (7.7%). Due to the high proportion of potassium feldspar cementation between quartz grains, a large number of corrosion pores are bound to appear in potassium feldspar during long-term immersion. Through microstructural observations, Zang et al. (1996) and Sun et al. (2022) confirmed that feldspar corrosion and calcite dissolution promote the development of pore structure in sandstones..

In summary, it can be concluded that the mechanism of water-rock interaction during long-term immersion is closely related to swelling of clay minerals and corrosion of potassium feldspar. Figure 4.13 illustrates schematically the effect of long-term soaking on the pore structure of sandstone and the mode of microscopic fractures. For dry sandstone, tensile damage tends to occur under compressive stress (Fig. 4.12), which is consistent with results of AE monitoring. The reason for this is mainly that the tensile strength of dry sandstone is significantly less than the compressive strength, and the quartz and cement are more prone to splitting during loading (Fig. 4.13b) (Xie et al., 2018b). Reviron et al. (2009) found that quartz grains do not weaken during long-term water immersion. However, the swelling of clay minerals leads to a change in the connection between crystal layers and an increase in interlayer spacing, and the corrosion of potassium feldspar promotes an increase in pore number and pore size (Fig. 4.13a). The above physicochemical reactions reduce the cementation action between the quartz particles. During loading, microcracks in long-term immersed sandstones are more likely to develop along weakened cementation surfaces. In addition, corrosion pores formed by the potassium feldspar provide space for the rotation and movement of the quartz grains, making them more prone to shear slip (Fig. 4.13b), thus leading to an increase in the number of shear cracks. This also explains the decrease in AE signals and UCS with increasing immersion time.

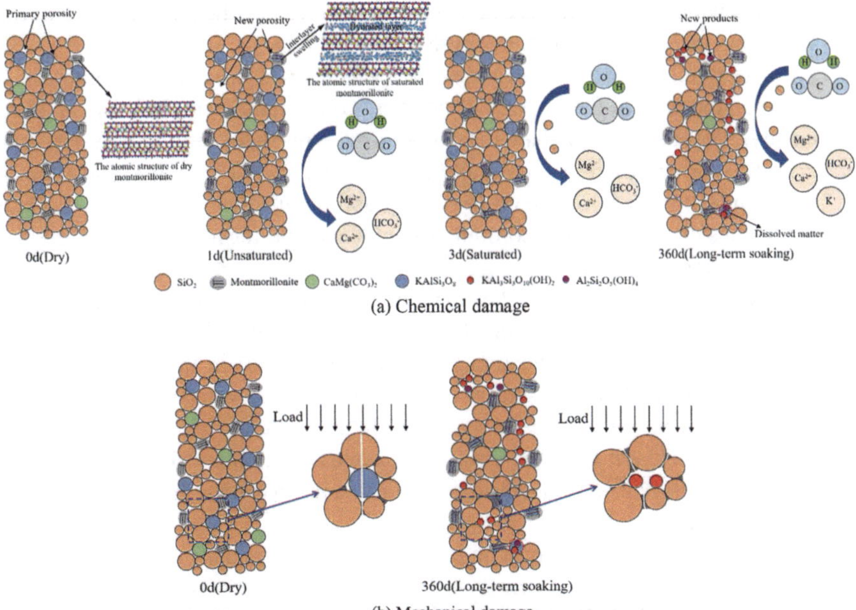


**Fig. 4.13** The effect of long-term immersion on pore structure of sandstone and mode of microscopic fractures. (**a**) Chemical damage, (**b**) Mechanical damage

## 4.5 Conclusions

In this study, based on a series of experimental characterizations (AE, XRD, SEM, NMR, wave velocity test, and hydrochemistry analysis), uniaxial compression experiments are conducted on sandstones with immersion times of 0d, 3d, 30d, 90d, 180d and 360d, respectively. The effect of immersion time on the pore structure, mechanical properties, and fracture development form of sandstone is studied, and finally the weakening mechanism of water-rock interaction during long-term immersion is discussed. The conclusions of the study are as follows.

1. Various physicochemical parameters (P-wave velocity, pore distribution, pH value, and ion concentration) directly or indirectly characterize the evolution of the pore structure in a sandstone. As the soaking time increases, changes in the pore structure of a sandstone can be divided into two stages, namely the swelling stage of clay minerals (saturation process) and the corrosion stage of potassium feldspar (long-term immersion process).
2. The UCS and AE signals of a sandstone decrease with increasing immersion time. Based on the relationship between stress and AE signals, the stress-strain curve can be divided into five phases. With the increase of immersion time, the stage ratio in elastic phase and stable fracture development phase of a sandstone gradually decreases, while phase ratio in compression phase and unstable frac-

ture extension phase gradually increases, indicating that the plastic behavior of a sandstone increases during the long-term immersion.

3. This study reflects the failure mode of microcracks during uniaxial compression based on the ratio of RA to AF. For dry sandstones, tensile fractures are approximately 2.3 times greater than shear fractures. With increasing immersion time, the ratio of shear fractures gradually increases from 29.1% in dry sandstones to 66.3% in sandstones immersed for 360 d, and the macroscopic failure mode of sandstones changes from a combination of shear and tensile crack surfaces to a single shear crack surface.
4. Hydrochemistry tests show the TDS in the soaking solution increases with increasing immersion time. The $HCO_3^-$ produced by the dissolution reaction of calcite and dolomite in sandstones leads to an increase in the pH value of the soaking solution. After the saturation stage, the corrosion reaction of potassium feldspar cements causes the pH value and $K^+$ concentration in the soaking solution to first increase, then decrease, and eventually remain stable. The variation of the pH value and $K^+$ concentration can provide a basis for the division of pore structure stages.
5. Based on the results of physicochemical experiments and mechanical tests, it can be assumed that microcracks in long-term immersion sandstones are more prone to developing along weakened cementation surfaces. Under the effect of stress, corrosion pores formed by the potassium feldspar provide space for the rotation and movement of quartz grains, making them more prone to shear slip. This is the microscopic reason for the change in failure mode and also explains the decrease in AE signals and UCS with increasing immersion time.

# Chapter 5
# Fluid-Solid Coupling Simulation Method for Water Immersion Weakening in Coal Mining

## 5.1 Introduction

The water-rock interaction (WRI) is one of the research focuses in the field of geology and geotechnical engineering. Many physical and chemical reactions are involved in the process of WRI, mainly including lubrication, precipitation, oxidation-reduction, ion-exchange, etc. Many geological hazards and engineering safety problems relate to WRI, e.g., slope stability (Zhao et al., 2018d), reservoir dam stability (Ukpai, 2021), rock bursting (Chen et al., 2019a), karst collapses (Bai et al., 2013), and water inrush (Huang et al., 2016; Li et al., 2019b; Ma et al., 2022a). The influence of water on the mechanical and seepage behavior of rock mass is the basis for the analysis and interpretation of the problems mentioned above. In recent years, a large body of research has investigated the effect of water on rock mechanical characteristics. It is generally found that the presence of water weakens the mechanical parameters (e.g., elastic modulus, compressive strength, cohesion, and tensile strength) as well as the brittle property of a rock (Baud et al., 2000; Erguler & Ulusay, 2009; Zhou et al., 2017; Talesnick & Shehadeh, 2007), and also changes the fragmentation distribution after rock failure (Haberfield & Johnston, 1990; Shen et al., 2020; Guha Roy et al., 2017; Kataoka et al., 2015). On the one hand, the existence of water increases the pore pressure in soil and rock mass, which reduces the effective stress of skeleton particles in the medium, so as to change the physical and mechanical parameters of the medium. On the other hand, the water may dissolve mineral compositions and cement between grains, so as to produce new mineral compositions. If the water contains some corrosive mineral components, the WRI would be more serious (Luo et al., 2021). For example, rocks with higher clay mineral contents are more susceptible to water (Verstrynge et al., 2014).

There are many aquifers within coal-bearing strata, and coal mining activities disturb these aquifers. Consequently, water inevitably flows into areas affected by coal mining, threatening the safe extraction of coal and even triggering water inrush incidents. In addition, water infiltration further weakens the coal rock mass, leading



C. Zhang, *Water Rock Interaction in Underground Coal Mining*,
https://doi.org/10.1007/978-981-95-9957-8_5

to frequent instability accidents in roadways, coal pillars, and other structures (Zhang et al., 2023c). With advancements in testing technologies and microscopic characterization techniques, substantial research has been conducted on the mechanical properties of coal-rock interaction with water and the microscopic mechanisms of water-rock interaction (Zhang et al., 2023b). However, most current studies focus on the laboratory scale, and the cross-scale effects of rock strength and differences in boundary environments prevent many research results from being directly applied to engineering site problems. Thus, achieving the scale-up application of laboratory results to engineering practice is critical. With the development of computer hardware and software, numerical simulations based on laboratory test results are becoming the primary research method for analyzing engineering-scale coal mining issues. However, previous studies mainly focused on the effects of in situ stress, mining-induced stress, mining techniques and parameters, stratigraphic characteristics, and geological structures on the stability of surrounding rock (Zhang et al., 2022g, 2023a; Xu et al., 2020, 2022), with limited attention given to the influence of water immersion weakening during coal mining. Moreover, in engineering-scale fluid-solid coupling simulations, most studies primarily focus on the interaction between the seepage field and the mechanical field, rarely considering the weakening effect of water on coal-rock masses (Li et al., 2023; Zhao et al., 2020c). For convenience, some studies addressing water immersion weakening omit seepage field analysis. For example, Poulsen et al. (2014) studied the effect of water immersion on coal pillar stability based on the mechanical properties of coal under dry and saturated conditions but did not consider the impact of moisture content and water seepage in their simulations. Similarly, Bai et al. (2016) analyzed the deformation mechanism of water-rich roadway roofs based on the relationship between immersion time and coal rock mass strength. However, this approach neither accounted for the influence of water seepage nor aligned simulation time with actual time effectively.

Based on this, this chapter proposes a fluid-solid coupling simulation method for water immersion softening, suitable for engineering scale, using the commonly used discrete element method. It also discusses the differences in its implementation within finite element simulation methods. Based on the proposed simulation method, the applicability, advantages, and disadvantages are discussed, providing a foundation for future research.

## 5.2 Fluid-Solid Coupling Softening Method in Finite Element Analysis

When the mine water comes into contact with the coal rock mass of the coal mine, the water-rock interaction will further weaken the rock mass; in this chapter, we take the stability of residual coal pillars (RCP) in contact with water as the engineering background. Therefore, a numerical method is essential to describe this fluid-solid coupling weakening process for the engineering-scale problem.

### 5.2.1 Permeability Calculation Model

Fracture networks in coal seams are anisotropic, resulting in permeability anisotropy. To describe the permeability anisotropy and its stress-dependent laboratory tests, which share similar stress environments with those of the coal pillar during the longwall retreating, were conducted to obtain the axial and radial permeability of the coal samples by Zhang et al. (2019h), and the schematic diagram of the test is shown in Fig. 5.1. The corresponding permeability can be expressed as follows:

$$\begin{cases} k_A = 7.9549e^{-4.8879\left(1-e^{-0.3888\sigma_h}\right)} \\ k_R = 13.1875e^{-6.8754\left(1-e^{-0.1324\sigma_V}\right)} \end{cases} \tag{5.1}$$

Where subscripts A and R represent the axial and radial permeability, respectively; $\sigma_h$ and $\sigma_v$ are the horizontal and vertical effective stress (MPa), respectively. It is obvious that the radial permeability is dependent on the vertical stress, while the axial permeability is mainly dependent on the horizontal stress (Zhang et al., 2019i).

In addition, for the RCP, it is evident from the field measurement that there were mainly elastic and plastic zones in the RCP, which also evolve during the longwall mining. Previous research shows that the permeability of coal increases

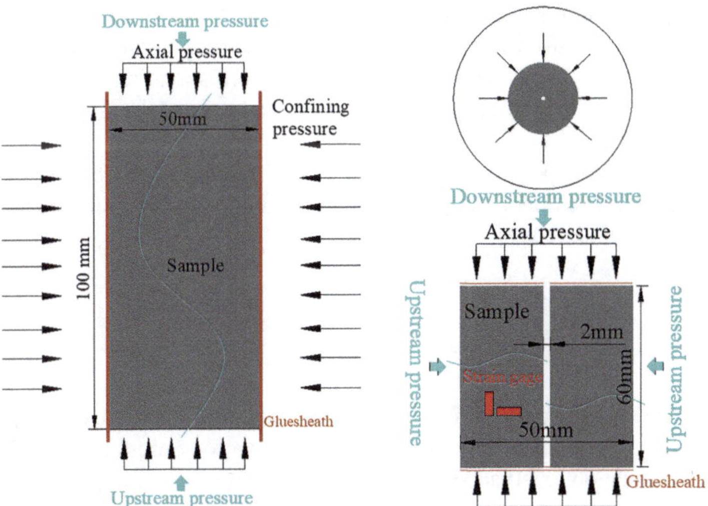


**Fig. 5.1** Axial (**a**) and radial (**b**) permeability test

significantly in the plastic zone. Therefore, it is necessary to distinguish the permeability variation in the plastic and elastic zones (Eq. 5.1). Equation (5.1) can be the permeability equation in the elastic phase because the tests were performed in the elastic stage. According to our previous work, the permeability in plastic coal samples can be expressed as Eq. (5.2).

$$\begin{cases} k_{s4} = 142.2316e^{-2.6349\left(1-e^{-0.3415\sigma_h}\right)} \\ k_{sR} = 142.2316e^{-2.6349\left(1-e^{-0.3415\sigma_v}\right)} \end{cases} \tag{5.2}$$

Where subscript S represents the permeability of the plastic coal sample.

### 5.2.2 *Water Invasion Weakening*

The underground reservoir generally remains stable after the impoundment of the goaf. Therefore, the overburden stresses on the RCP bearing remain stable during water storage. During this period, the stability of the RCP mainly depends on the water immersion softening (Yao et al., 2020a; Zhang et al., 2023a; Han et al., 2023). Thus, a fluid-solid coupling softening method was proposed here to describe this degradation. Considering the computational efficiency, this study makes the following assumptions: (1) The study area belongs to a low-gas coal seam; the seepage in the coal pillar is saturated water seepage without gas. The equations and formulation for saturated fluid flow can be found in FLAC3D 6.0 documentation (Itasca Consulting Group, 2019); (2) When the pore pressure in the RCP is greater than 0, the coal begins to be immersed in water; (3) The slow-flow velocities in the pores obey Darcy's law, and the effects of temperature are negligible. Darcy's law can be used to describe the low Reynolds number fluid in porous media. The upper limit of the Reynolds number corresponding to Darcy's law is between 1 and 10 (Du et al., 2019a, 2022b). The velocity of the coal pillar belongs to low-speed flow.

Permeability was updated after the mechanical calculation according to the permeability model. The relationship between permeability and porosity is given in Eq. (5.3) (Koponen et al., 1997).

$$\frac{k}{k_0} = \left(\frac{\varphi}{\varphi_0}\right)^3 \tag{5.3}$$

Where $k$ and $k_0$ are the permeabilities before and after deformation, $\varphi$ and $\varphi_0$ are the corresponding porosities. Thus, the porosity under various stress and elastic-plastic conditions can be obtained according to Eq. (5.3). The change of porosity will further cause the change of moisture content.

In addition, the water invasion weakening of the coal matrix in RCP is a time-dependent process. However, few studies have been reported on the influence of long-term water immersion (years or even decades) on coal strength. Existing works show that the coal strength slightly reduces after the coal matrix is immersed in water for a certain time (usually less than 1 week) (Yao et al., 2019b). This is mainly due to the correlation between immersion time and moisture content (Fig. 5.2). The water content exponentially increases with the immersion time; as a consequence, the strength gradually degrades. Therefore, in the simulation, the effect of time-dependent water content should also be considered.

Due to the complex stress state and its constantly changing nature, the porosity of the RCP varies during the entire process, suggesting that the maximum moisture content of the coal matrix is different. In particular, the porosity increases as the coal matrix yields. Therefore, it is necessary to calculate the maximum water content first. According to the first assumption, the maximum moisture content ($\omega_m$) of the coal is:

$$\omega_{\mathrm{m}} = \frac{\rho_{\mathrm{w}}\varphi}{\rho_{\mathrm{w}}\varphi + \rho_{\mathrm{m}}} \tag{5.4}$$

Where $\rho_m$ and $\rho_w$ are the densities of dry coal samples and water, respectively, in Kg/m$^3$. And then, the real-time moisture content was updated according to Eq. (5.5) in the mechanical calculation.

$$\omega = \omega_{\mathrm{m}}\left(1 - \exp\left(-0.077t\right)\right) \tag{5.5}$$

Where $t$ is the immersion time, and it begins when the pore water pressure of the zone is greater than 0 according to the second assumption.

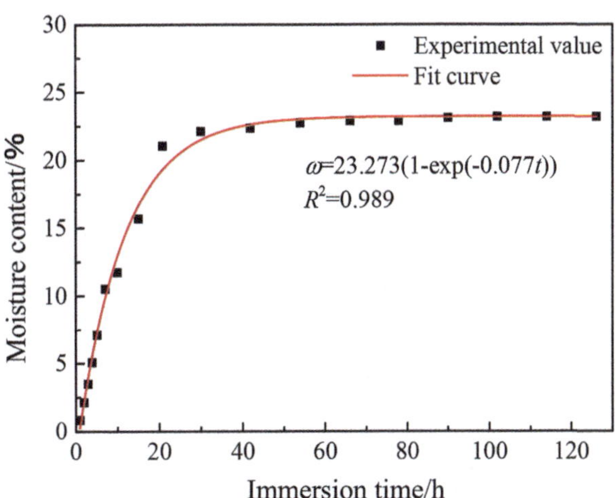


**Fig. 5.2** Coal samples' moisture content with different immersion time (Modified after Yao et al., 2019b)

Combining Eqs. (5.1), (5.2), (5.3), (5.4), and (5.5), the permeability and moisture content for each zone in the numerical model can be calculated based on its stress regime and stress state (elastic or plastic) different elastic and plastic zone stress states. The strength of the coal sample is closely related to its moisture content, which can be obtained by laboratory tests. Yao et al. (2019b, 2020a) tested the effect of moisture content on the uniaxial compressive strength and shear strength of the coal seam in our study mining area, and found that the elastic modulus, cohesion, and internal friction angle decrease exponentially with water content (Fig. 5.3).

Figure 5.3 shows the change curve of the mechanical parameters of coal samples with the increase in moisture content. Among them, the elastic modulus of coal samples is obtained by uniaxial compression tests, and the cohesion and internal friction angle are obtained by compression shear tests. The relationship between elastic modulus ($E$), cohesion ($c$), internal friction angle ($\Phi$), and water content can be expressed by an exponential function.

$$\begin{cases} \Phi = 0.0603\exp(-0.17\omega) + 0.9397 \\ c = 0.2943\exp(-0.12\omega) + 0.7057 \\ E = 0.5344\exp(-0.34\omega) + 0.4656 \end{cases} \tag{5.6}$$

Considering the above factors, the indirect fluid-solid coupling simulation method is adopted in this chapter. Alternate circulation of mechanical and seepage calculation is conducted. In mechanical calculation, the state (plastic or elastic) and stress are obtained, and permeability, porosity, and maximum moisture content are updated accordingly. In seepage calculation, pore pressure and real-time moisture content are obtained, and coal pillar strength is updated accordingly. The detailed process is shown in Fig. 5.4.

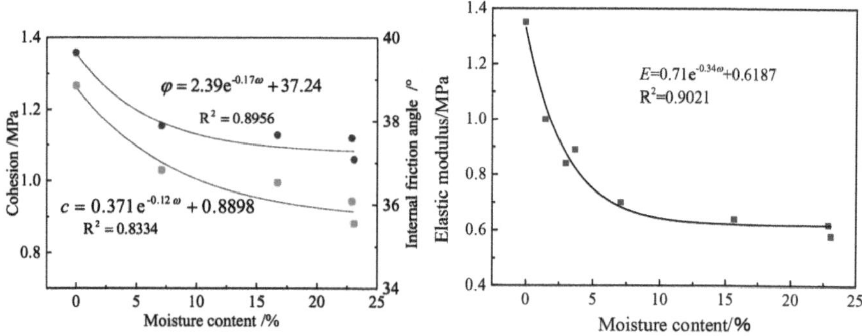


**Fig. 5.3** Elastic modulus, internal friction angle, and cohesion as a function of water content. (Modified after Yao, et al., 2019b, 2020a)

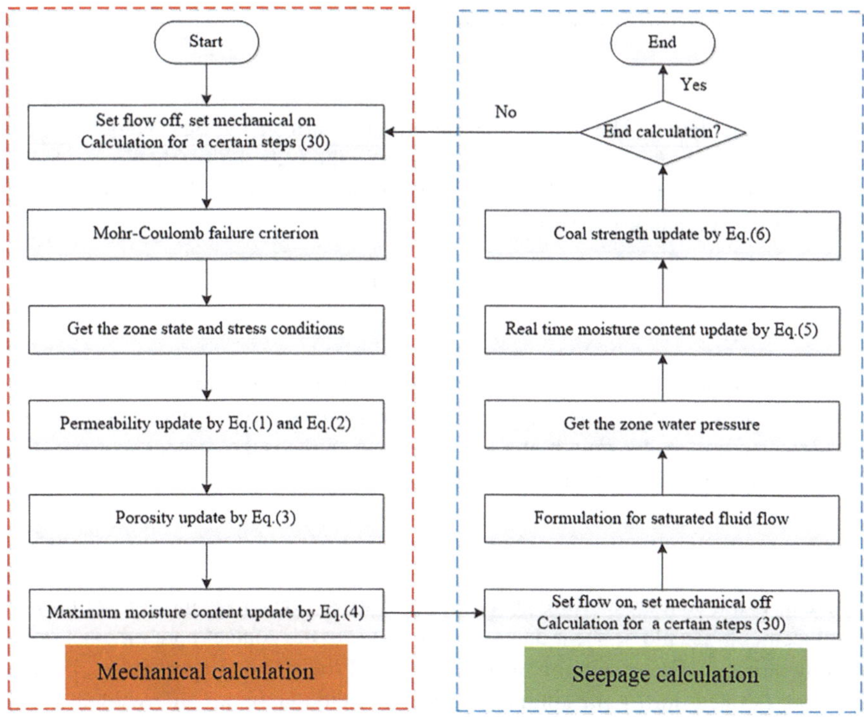


**Fig. 5.4** Main procedure of fluid-solid coupling in water immersion softening

## 5.3 Fluid-Solid Coupling Softening Method in Discrete Element Analysis

Discrete element numerical simulation can effectively simulate fractured rock masses; the discrete element simulation method in this chapter is mainly based on the engineering background of roadway stability under the influence of water-filled fault structures. Due to the existence of roof aquifers, two aspects need to be considered when exposing faults in roadways: fault activation and water invasion weakening. When fault structures are activated, a seepage channel of the roof aquifer has been formed. The roadway surrounding rock (RSR) strength decreases because of the influence of water immersion weakening, which eventually leads to the instability of the surrounding rock. To elaborate the effect of water immersion weakening on the progressive failure of the RSR, a numerical simulation method of fluid-solid coupling weakening on the engineering scale is established.

### 5.3.1 Permeability Calculation

The UDEC numerical simulation software can perform a coupled mechanical-hydraulic calculation. The mechanical deformation affects the flow velocity of the fluid, and the pore pressure of the contact also influences the mechanical calculation. The calculation of groundwater seepage through contacts is based on the cubic law (Moon & Fernandez, 2010):

$$q = -K_{\mathrm{j}} a^3 \frac{\Delta P}{l} \tag{5.7}$$

where $q$ is the seepage rate of a single contact, $K_{\mathrm{j}}$ is a contact permeability factor, $\mu$ is a dynamic viscosity of the fluid, $\Delta P$ is a pressure difference between upstream and downstream, $l$ is the length of a single contact, and $a$ is the contact aperture. The hydraulic aperture is given, in general, by:

$$a = a_0 + \Delta a \tag{5.8}$$

where $a_0$ is the contact aperture at zero normal stress, and $\Delta a$ is the contact normal displacement. Tensile or shear damage may occur in the contact under the action of mining stresses. The maximum aperture $a_{\mathrm{max}}$ and the residual aperture $a_{\mathrm{res}}$ in contacts are defined during the simulation to obtain the maximum and residual water content of coal and rock mass. For the case of irregular polygons inside the model, the permeability coefficient of the model can be obtained by simplifying Eq. (5.7):

$$k_{\mathrm{f}} = \frac{\rho_1 g a^2}{12\mu} \tag{5.9}$$

where $\rho_1$ is the fluid density, $k_{\mathrm{f}}$ is the permeability coefficient, and $\Sigma l$ is the sum of the contact lengths inside the model. In isotropic media, the relationship between the permeability coefficient and permeability is: (Zhang et al., 2019h)

$$k_{\mathrm{f}} = \frac{k \rho_1 g}{\mu} \tag{5.10}$$

where $g$ is the acceleration of gravity and $k$ is the permeability of porous media. In the UDEC numerical simulation, the influence of mechanical deformation on seepage can be realized by using the relationship between contact permeability and contact aperture. Combined with Eqs. (5.8), (5.9), and (5.10), the relationship between the contact aperture and permeability is as follows:

$$k = \frac{(a_0 + \Delta a)^2}{12} \tag{5.11}$$

### 5.3.2 Water Invasion Weakening Criteria

A 2D Voronoi model consists of unit blocks and contacts between blocks. In the UDEC simulation, it is assumed that the seepage process only occurs in the cracks between the blocks. After roadway excavation, the stress state of RSR changes, and the block exhibits motion forms such as rotation, movement, and deformation, while the contact produces the failure form of shear and tension. Figure 5.5c shows that a contact of tension and shear will form a seepage channel for water in the UDEC simulation. Assuming that the density of the block and the pore pressure of the roof aquifer remain constant, the water seepage in strata depends on the contact aperture and permeability. Figure 5.5 shows that the change in moisture content in coal and rock is related to the contact aperture.

To effectively count the variation law of moisture content of coal and rock masses, the coal and rock mass in the study area is divided into a 1 × 1 m rectangular grid, as shown in Fig. 5.5a. The moisture content of units is as follows:

$$\eta = \frac{\rho_1 \Sigma la}{\rho_1 \Sigma la + \rho_2 \Sigma V} \tag{5.12}$$

where $\rho_2$ is the density of blocks in the unit, and $\Sigma V$ is the volume of blocks in the unit.

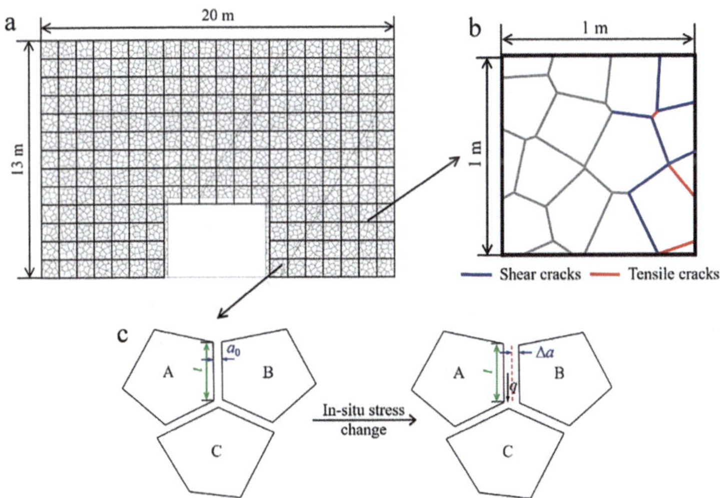


**Fig. 5.5** (**a**) Unit division, (**b**) a schematic diagram of damage ratio, and (**c**) the representation method of moisture content

### 5.3.3 Weakening Coefficient Acquisition

Through the uniaxial compression tests and compression-shear tests of coal and rock samples under different moisture content conditions, the weakening relationship between moisture content and mechanical parameters in the numerical simulation is obtained, as shown in Fig. 5.6. The experimental results show that the weakening coefficients of mechanical parameters of coal and rock samples are negatively correlated with the moisture content. When the moisture content of sandstone changes from 0% (dry state) to 5.05% (saturated state), the weakening coefficients of elastic modulus and cohesion of sandstone change greatly, decreasing by 55.7% and 62.6%, respectively. The variation trend of the weakening coefficients of the mechanical parameters of coal samples is consistent with that of the sandstone. Therefore, the above water invasion weakening relationship can be embedded in the UDEC numerical simulation software through the fish language, and the fluid-structure coupling analysis of the RSR stability under the water invasion weakening condition can be realized.

However, the relation between water contents and mechanical parameters in Fig. 5.6 is applicable only to the strength weakening results of intact rock masses, while the relationship between moisture content and residual strength cannot be reflected. Therefore, in the proposed algorithm, the coal and rock masses are assumed to have been damaged when the contact within rock masses produces tensile and shear cracks. The mechanical strength parameters of contacts in units will be given residual values.

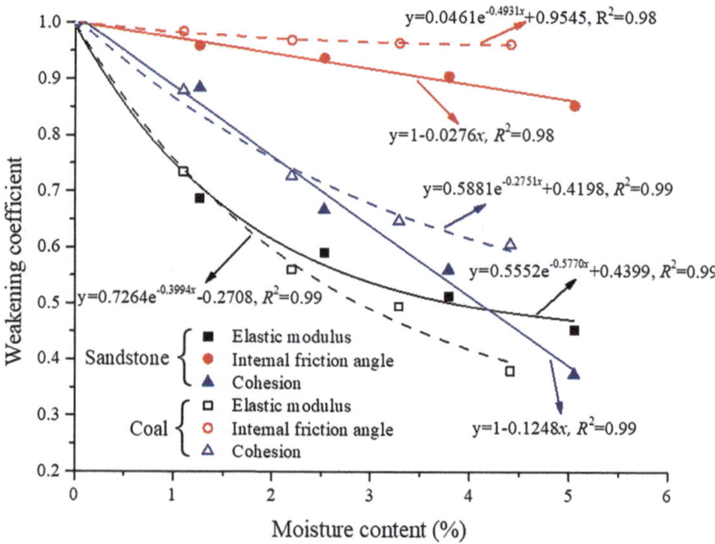


**Fig. 5.6** Variation of the weakening coefficient of coal and sandstone with different water contents

Figure 5.7 shows the numerical simulation program of RSR stability under the condition of water invasion weakening. After the excavation of the roadway is completed, the mechanical calculation is 100 steps, and then the height of the fracture zone $l_s$ (shear cracks or tensile cracks) around the roadway is checked to see whether it develops to a roof aquifer. A further 100 calculation steps are run if cracks in the model do not develop to the aquifer. When the cracks have developed to the aquifer, the model carries out 100 steps of seepage calculation, and the moisture content in units is determined by Eq. (5.12). When the pore water pressure in units is greater than 0 MPa, the assignment of mechanical parameters of contacts is negatively correlated with moisture content (Fig. 5.6). According to Eq. (5.11), the permeability coefficient in units raises with increasing contact aperture. When tensile or shear cracks occur in units, the mechanical parameters and seepage parameters of contacts in units will be given the residual value and the maximum value, respectively, and then the model will carry out 100 mechanical calculation cycles. The iterative process proceeds until the roof displacement of the roadway is consistent with field tests.

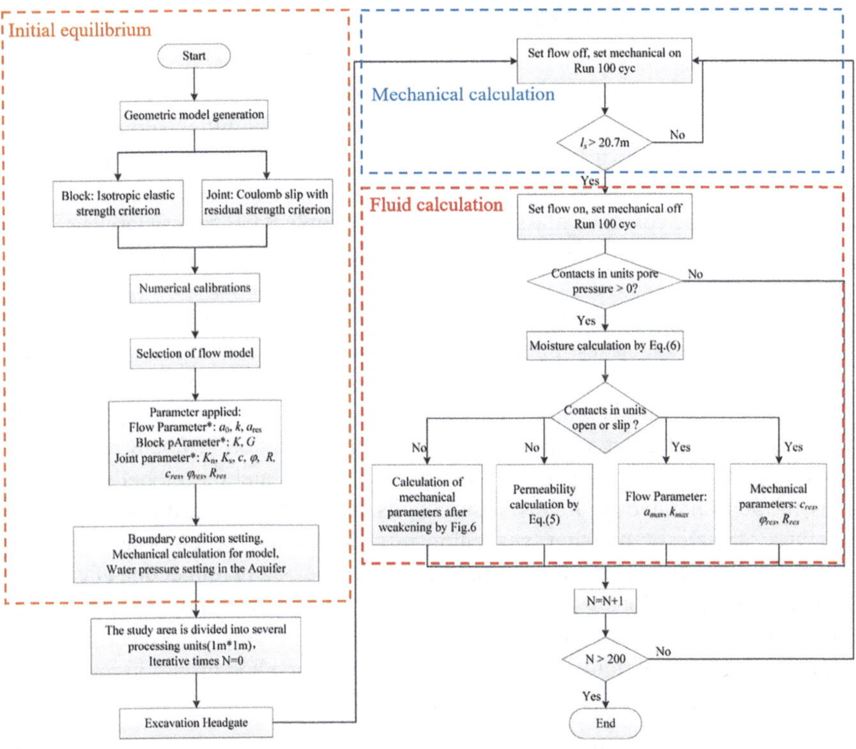


**Fig. 5.7** Numerical simulation program under water invasion weakening

### 5.3.4 Damage Ratio Calculation

Due to the redistribution of surrounding rock stress after roadway excavation, cracks will occur in the RSR. In previous numerical simulation analyses, it was hard to obtain the damage to the surrounding rock in a specific region only by the crack distribution map. To accurately obtain the degree of failure of the RSR after excavation, an index of the damage ratio of the RSR is proposed in this chapter. The damage ratio refers to the ratio of failure contacts (tension or shear) of the monitoring area (or the whole model) to all contacts in the monitoring area during roadway excavation. For the convenience of statistics, the monitoring unit of the surrounding rock damage is a rectangular grid with a size of 1 × 1 m (Fig. 5.5b), and the unit damage ratio is as follows:

$$D = \frac{\Sigma l_i}{\Sigma l} \tag{5.13}$$

where $D$ is the unit damage ratio, and $\Sigma l_i$ is the sum of crack lengths in the unit.

## 5.4 Discussion and Outlook

The water immersion weakening effect of coal rock mass is an important factor leading to the instability of surrounding rock during coal mining. Numerous laboratory studies have shown that the water-rock interaction significantly weakens the strength of coal rock mass, and a quantitative relationship between the mechanical parameters of the coal rock mass and its moisture content has been established. How to apply the quantified results obtained from laboratory tests to field engineering practice is crucial. Based on this, this chapter proposes a fluid-solid coupling simulation method that takes into account water immersion softening. The method addresses two main issues: First, the permeability is updated in real time based on mining-induced stress and the yield state of the coal rock mass. Second, the strength of the coal rock mass is updated in real time according to its moisture content. On this basis, the softening characteristics of water immersion in fractured structural regions of tunnels were analyzed through simulation, and the model's feasibility was further validated by field measurements. In addition to being applied to the analysis of water immersion weakening in coal mining, the proposed fluid-solid coupling simulation method is also applicable to other water immersion engineering problems, such as slope stability under rainfall, stability of reservoir dams, and stability analysis of tunnels and chambers in water-rich strata. However, the water immersion softening fluid-solid coupling simulation method proposed in this chapter still has areas that require improvement.

1. The fluid-solid coupling simulation method for water immersion softening proposed in this chapter adopts an indirect fluid-solid coupling approach. After a

fixed calculation step, the seepage and mechanical calculations are separated and iteratively coupled. The advantage of this method is its relative simplicity and efficiency, with a small computational load, making it suitable for numerical simulations at an engineering scale. However, in practical scenarios, seepage and mechanical calculations occur simultaneously, which makes it crucial—but difficult—to determine reasonable time intervals when performing indirect fluid-solid coupling calculations. For large-scale engineering, especially in coal mining, where mining-induced damage is significant and the scale is vast, achieving fully coupled calculations is challenging.
2. In this chapter, the water immersion weakening is realized using the relationship between moisture content and the strength of the coal rock mass. However, in reality, in addition to moisture content, factors such as the number of wet-dry cycles, immersion time, and seepage water pressure also affect the strength of the coal rock mass. For coal mining, long-term water immersion more closely reflects the actual conditions on-site regarding the weakening of coal rock mass strength. However, using immersion time to model the weakening of coal rock mass strength presents two main issues. On the one hand, experimental results on long-term water immersion (over 1 year) are relatively scarce, and there is little available quantitative relationship between coal rock mass strength and immersion time. On the other hand, calibration of time in numerical simulations is not straightforward. Currently, calibration of simulation time mainly relies on comparing field measurements with simulation results. However, the simulation computation time is influenced by model size and mesh discretization, making it difficult to align with actual time.
3. The water immersion softening fluid-solid coupling simulation method proposed in this chapter helps to apply laboratory results to an engineering scale. However, the strength of the coal rock mass and seepage characteristics themselves exhibit scale effects. Whether the weakening coefficient obtained in the laboratory is consistent with the engineering weakening coefficient requires further confirmation. Particularly for the seepage characteristics of the coal rock mass, as the scale increases, the continuous faces of the coal rock mass will also increase, which will gradually enhance its seepage capacity. Therefore, the permeability update model and the mechanical parameter weakening model obtained from laboratory tests need to undergo scale verification. In addition, due to the differences in the coal rock mass's depositional environment, its mechanical and seepage properties have engineering-specific characteristics. For specific coal mining engineering problems, targeted water immersion weakening experiments and permeability stress sensitivity tests are required to obtain the updated formulas for mechanical parameters and permeability in the simulation process.
4. The fluid-solid coupling simulation method considering water immersion weakening proposed in this chapter can be applied to both discrete element and finite element simulation methods and is well-suited for commonly used coal mining numerical simulation software, such as UDEC, PFC, and FLAC3D. With the development of simulation methods, many software programs now support coupled simulations of finite element and discrete element methods (finite-

discrete element method), eliminating the need to predefine fracture meshes as in discrete element methods. Instead, fracture elements can be randomly generated based on failure criteria. Therefore, future work can further expand the existing numerical simulation method considering water immersion weakening to the finite-discrete element method.

# Chapter 6
# Failure Analysis of Residual Coal Pillar Under the Coupling of Mining Stress and Water Immersion

## 6.1 Introduction

Arid and semi-arid areas in northwestern China have become the main coal production areas. Among these areas, the total proven reserves in Shaanxi, Inner Mongolia, Ningxia, Gansu, and Xinjiang have reached 1.0628 trillion .tons, approximately accounting for 81% of the total proven reserves of the country. In 2020, coal production in these regions accounted for 86.1% of the country's total coal yield (National Bureau of Statistics, 2021). However, these five provinces have a serious shortage of water resources, accounting for only 1.6% of the country's total water resources, and the surface ecosystem is fragile (Wang et al., 2018).

In recent years, the movement of rock stratum caused by large-scale and high-intensity coal mining has led to the influx of large-scale shallow water into the underground coal mine (Zhang et al., 2021a), and the ecological water resources have been further destroyed. The northwest five provinces belong to arid to semi-arid regions (Fig. 6.1a), where vegetation and ecological water levels are difficult to recover after coal mining, and water scarcity is further aggravated (Qiao et al., 2017; Xie et al., 2018a). As shown in Fig. 6.1a, b, the area with water depth less than 5 m accounts for 40.9% of the mining area before large-scale mining in the Yushenfu mining area, and the water depth decreased by more than 8 m in the area of about 758.9 km$^2$ (7.3% of the mining area) after coal mining (Fan et al., 2016). Therefore, the protection and utilization of water resources is an important scientific and technological problem to be solved in green coal mining in northwestern China.

The mining of coal seams will inevitably lead to the fracture of overburden. Traditional water-retaining methods (fill mining, strip mining, height-limited mining, etc.) (Peng & Chiang, 1984; Renani & Martin, 2018) avoid the development of a water-conducting fracture zone in the longwall face to aquifers by increasing the support capacity of the goaf. However, the specific water isolation structure is the premise of water-retaining methods, which determines the success of water-retaining methods, and the traditional water-retaining method technology has some



C. Zhang, *Water Rock Interaction in Underground Coal Mining*,
https://doi.org/10.1007/978-981-95-9957-8_6

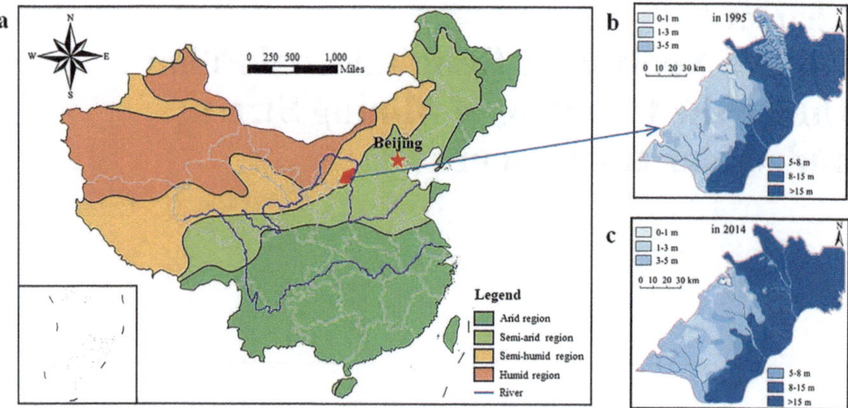


**Fig. 6.1** Water-level distribution map. (**a**) Distribution map of arid and semi-arid areas in China. (**b**) Contour map of depth to underground water before coal mining. (**c**) Contour map of depth to underground water after coal mining

problems, such as low efficiency, poor benefits, low resource recovery rates, and so on. Given the above problems, combined with the existing technical achievements (Gu, 2013; Pujades et al., 2016; Menéndez et al., 2019; Fan et al., 2020), Gu (2015) proposed the underground-reservoir, water-storage technology of "conduct-store-use" of mine water, which alleviated the contradiction between coal mining and water resource protection in the ecologically fragile mining area in western China.

In the course of the implementation of the underground reservoir, the coal pillar used to isolate the goaf and longwall face will be under the effect of mining-water invasion for a long time. In the process of mining in the longwall face and water storage in the goaf, the coal pillar in the complex stress and water environment is superimposed by dynamic and static loads, such as broken roof lateral abutment pressure, overlying strata pressure, goaf water pressure, and mine earthquake impacts (caused by large-scale subsidence of room-type goaf or large-scale collapse of the thick hard roof in mining) (Wang et al., 2016b; Bai et al., 2017; Zhang et al., 2019c), as shown in Fig. 6.2. In the process of long-term water invasion to coal pillars, groundwater changes the mineral composition and microstructure of the coal through physical and chemical actions, such as lubrication, softening, erosion and transport, disintegration, mudding, strengthening of combined water, ion exchange, dissolution, and redox (Maruvanchery & Kim, 2019), which increases its porosity and causes internal damage to the rock mass. Microcracks propagate to form macroscopic cracks, thus weakening the macroscopic mechanical properties of the coal (Eberhardt et al., 1999; Heggheim et al., 2005). Therefore, the stability of coal pillars under long-term mining-water invasion is one of the key factors in determining the stability of the ground reservoir (Salmi et al., 2017), and it is also an important factor affecting the safe production of the follow-up longwall face.

In recent decades, scholars have performed a great deal of research on coal-pillar stability and corresponding dimension design utilizing field measurements,

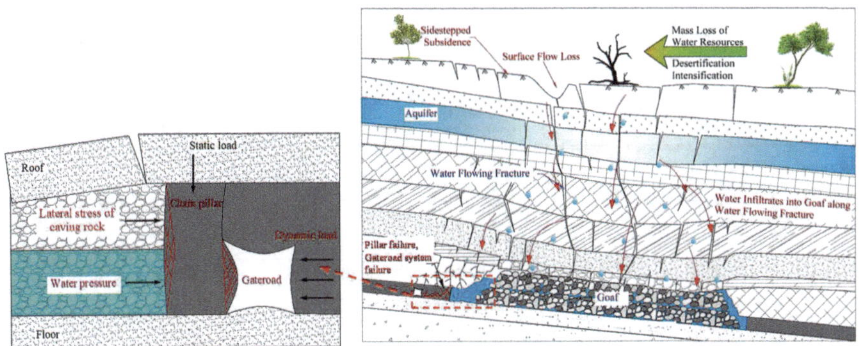


**Fig. 6.2** Eco-environmental damage effects caused by coal-seam mining

theoretical analysis, and numerical simulation and have achieved abundant research results (Hustrulid, 1976; Peng, 2013; Zhang et al., 2021b). However, most existing research results are based on the stability analysis of coal pillars during the normal production stage and rarely involve the coupling effects of long-term water invasion and mining. Poulsen et al. (2014) analyzed the strength reduction of saturated coal and coal measure rock and their influence on the strength of coal pillars through a numerical model of coal-pillar strength reduction, while the water movement in coal pillars was not considered, as well as mining disturbance. Wang et al. (2019) used UDEC to simulate the weakening effects from water invasion on coal pillar stability after coal mining was stabilized. Due to the two-dimensional simulation, the influence of longwall face mining on coal pillars could not be simulated.

Thus, stress and water are two key physical fields acting on coal pillars, and the coupling of dynamic and static loads and water invasion causes "damage-flow-weakening-damage" and progressive failure of coal pillars. However, most research at present on the stability of coal pillars from mining disturbance and water invasion is still at the laboratory scale (mostly coal samples), while the research on the coupling of mining stress and water invasion on coal pillars at the engineering scale is relatively less common. Based on this, the stability of coal pillars under the condition of mining-water invasion is studied by combining numerical simulations and field measurements at the engineering scale; the variation law of cumulative damage to coal pillars is obtained, and a reasonable method for coal-pillar size reservation is given.

## 6.2 Process of Pillar and Gate Road Failure

### *6.2.1 Engineering Geological Conditions*

The studied coal mine is located in northern Yulin City, Shaanxi Province, as shown in Fig. 6.3a. The main mining coal seam 2–2 is located in the Yan'an Formation. The dip angle of this coal seam is 0–5°, the thickness of the coal seam is 8.89–11.73 m,

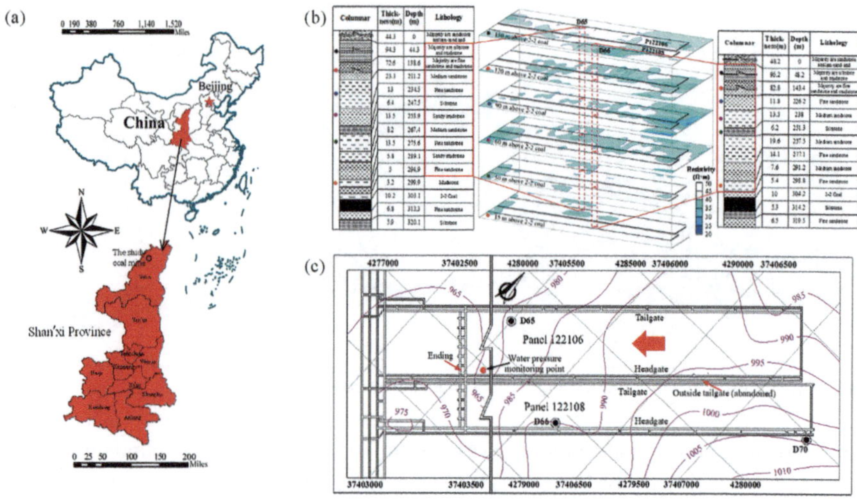


**Fig. 6.3** (**a**) Location of the study coal mine. (**b**) Generalized stratigraphic column and slice map of the ground transient electromagnetic anomaly area. (**c**) Longwall face layout

and the average thickness is 10.41 m. The Yan'an Formation bedrock fissure aquifer (15–60 m above the coal seam), Zhiluo Formation bedrock fissure aquifer (60–150 m above the coal seam), and weathered bedrock aquifer (150 m above the 2–2 coal seam) are located in the overburden above the coal seam, as shown in Fig. 6.3b. P122106 is the first longwall face of the study mine; the southern part of the P122106 longwall face is the P122108 longwall face; and the layout of the longwall face is shown in Fig. 6.3c. Among these, the P122106 longwall face has full seam mining, strike length of the P122106 longwall face is 6200 m, inclined length is 350 m, and mining height is 6 m. The P122108 longwall face has caving mining, inclined length is 280 m, and mining height is 10 m. Considering the four geological exploration bores in and near the two longwall faces, the immediate roof of the longwall face in the study area is composed of sandstone, and the main roof is composed of fine sandstone with an average thickness of 12.1 m and 14.2 m, respectively. The floor is composed of mudstone, siltstone, and fine sandstone with an average thickness of 12.3 m.

The tailgate and headgate are arranged in parallel on both sides of the longwall face. The size of the coal pillar is 35 m between the P122106 headgate and the P122108 tailgate, and a P122106 outside tailgate has been excavated between the two gate roads. The gate road layout at the study site is presented in Fig. 6.4. The cross section of the gate road is 5 × 3.9 m (width × height), and the net sectional area is 18.5 m$^2$. The roof of the gate road is supported by bolts and cables, and steel mesh and bolts are used on both sides of the gate road. After the P122106 longwall face was excavated to the ending line, the P122106 gate roads were completely closed, the P122106 longwall face was no longer drained, the goaf was transformed into a

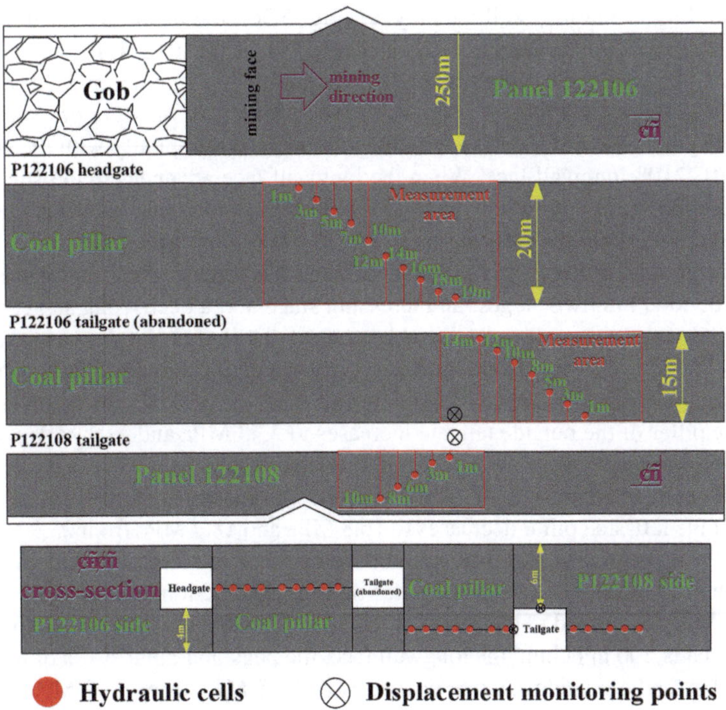


**Fig. 6.4** Layout of gate road and monitoring area

groundwater reservoir, and a water pressure monitoring point was constructed near the P122106 headgate, as shown in Fig. 6.3c.

## 6.2.2 *Instability Analysis of Gate Road and Coal Pillar*

Stress and displacement around the gate road have changed during the P122106 retreat. Hydraulic cells were installed in the coal pillar to monitor the stress variation process, and a surface displacement monitoring station was installed in the P122108 headgate roof and coal-pillar rib to record the rock mass movements. According to the actual needs and the mining situation of the longwall face, the measuring instruments were installed in the measurement area 150 m ahead of the P122106 longwall face, as shown in Fig. 6.4. Ten hydraulic cells were installed in the pillar rib of the P122106 headgate, and seven and five hydraulic cells were installed in the coal pillar and solid side of the P122108 tailgate, respectively. The diameter of the borehole is 48 mm, the height from the gate road floor is 1.75 m, and the horizontal distance between the adjacent cells is 3 m.

According to the longwall face layout (Fig. 6.3c), the position near the ending line of the P122106 headgate is the lowest point of the P122106 longwall face; therefore, the water pressure measured at this point is the maximum water pressure in the P122106 goaf. Figure 6.5 shows the variation curve of water pressure in the P122106 goaf. The water pressure in the goaf increases gradually with the advance of the P122108 longwall face. When the longwall face advances to the study area, the measured water pressure in the goaf is 0.06 MPa, which indicates that the maximum water height in the working face of P122106 is approximately 6 m.

In the process of longwall face advancement, the vertical stress in the coal pillar can be divided into two stages: an increasing stage and a decreasing stage, and the coal-pillar stress always presents a saddle-shaped distribution, as shown in Fig. 6.6a. Due to the action of the front abutment pressure of the longwall face, when the monitoring area is 40 m ahead of the longwall face, the peak abutment stress of the left-side pillar of the outside tailgate increases to 23.3 MPa and 21.1 MPa with distances of 4 m and 6 m to the left-side pillar ribs, respectively. When the longwall face continues to advance to 40 m ahead of the monitoring area, the peak abutment stress of the left-side pillar decreases to 16.8 MPa and 18.7 MPa from 23.3 MPa and 21.1 MPa, respectively, and its position develops 2 m and 2 m to the interior of the coal pillar, respectively, which indicates that the destruction area of the left pillar has gradually expanded to the center of the pillar from the two ribs. When the monitoring area is 200 m behind the longwall face, the peak abutment stress of the right-side pillar of the outside tailgate decreases to 17.5 MPa from 13.8 MPa, and the peak abutment stress of the P122108 tailgate solid rib increases from 13.8 MPa to 17.5 MPa, which indicates that the vertical stress in the left-side pillar gradually transfers to the right-side pillar and the solid side of the P122108 tailgate.

Figure 6.6b shows the deformation monitoring curve of the P122108 tailgate during the mining process of the P122106/P122108 longwall face. As shown in Fig. 6.6b, the displacement of the coal pillar and roof increases with the advancement of the P122106 longwall face, and the overall distribution is S-shaped. When

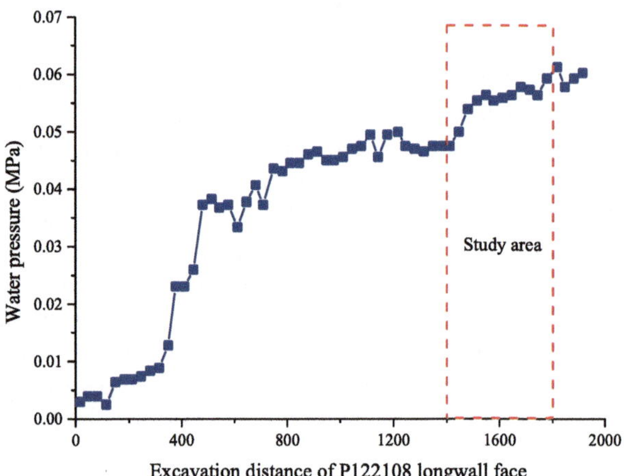


**Fig. 6.5** Water pressure change diagram at the water-level monitoring point

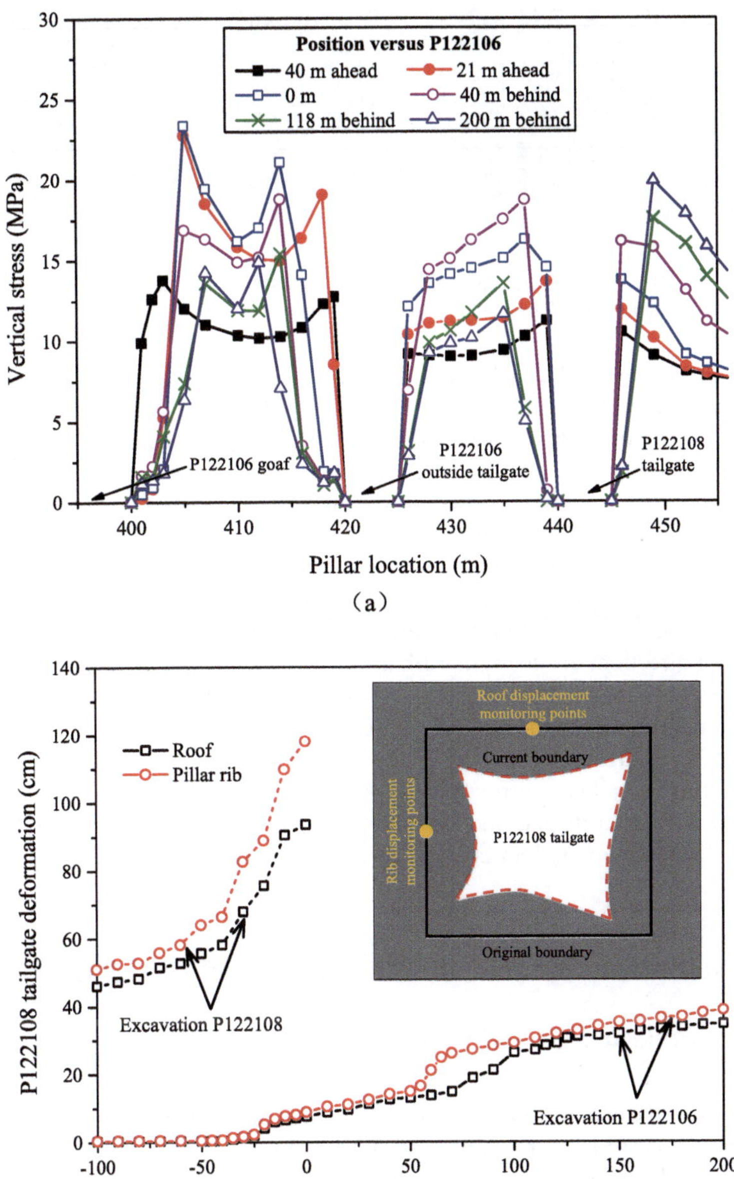


**Fig. 6.6** During the mining process of the P122106/P122108 longwall face. (**a**) Stress distribution characteristics of the coal pillar. (**b**) P122108 tailgate deformation

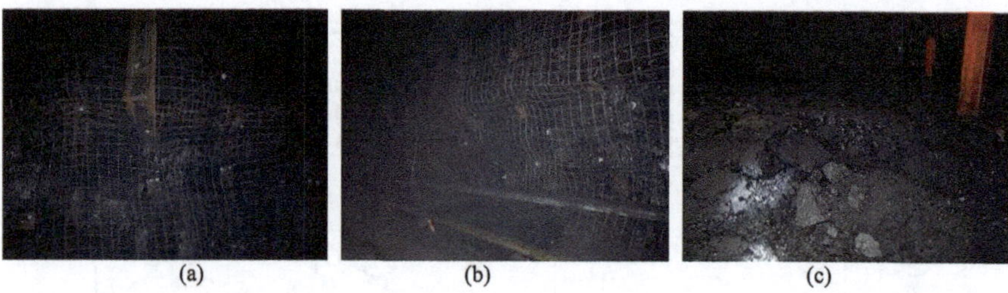


**Fig. 6.7** P122108 gate road deformation failure. (**a**) Roof-bending subsidence. (**b**) Large coal-pillar rib deformation. (**c**) Floor cracking and floor heave

the P122106 longwall face advances, the maximum deformation of the tailgate roof and pillar rib is 34.6 cm and 38.7 cm, respectively. When the P122108 longwall face is located approximately 40 m behind the monitoring area, the tailgate is affected by the front abutment stress of the longwall face, and the deformation of the tailgate surrounding rock increases sharply. Among these, the deformation of the coal-pillar rib is the most severe, and its deformation rate increases from 0.6 cm/m to 2.1 cm/m. When the P122108 longwall face advances to the monitoring area, the maximum deformation of the pillar rib and roof increases to 118.0 cm and 93.5 cm, respectively, and the deformation of the tailgate continues to increase. At this time, in the horizontal direction, the coal-pillar rib is crushed; the maximum depth of rib spalling can reach 2 m; steel mesh is seriously deformed; part of the bolt falls off; and water seepage occurs in the blot hole of the coal-pillar rib (Fig. 6.7b). In the vertical direction, the extrusion movement of the roof leads to serious bending of the steel beam. When the tailgate surrounding rock is equipped with blots and steel mesh to limit the horizontal displacement of the two ribs, the floor will produce lift over time, and the maximum floor heave is approximately 1 m, as shown in Fig. 6.7.

## 6.3 Numerical Simulation of the Stability of Coal Pillars

### *6.3.1 Numerical Model Construction*

According to the actual situation of the P122106 and P122108 longwall faces, the $X$ direction of the model is the slope of the longwall face, $Y$ direction is parallel to the strike of the longwall face, and $Z$ direction is perpendicular to the $XY$ plane. The model size is $X = 780$ m, $Y = 400$ m, and $Z = 88.6$ m, as shown in Fig. 6.8. Grid refinement in a monitoring area of a simulated coal pillar is used in this chapter, grid size is refined to 0.5 m, and the grid far away from the monitoring area is gradually increased to avoid stress discontinuity. In situ rock stresses were measured in four locations in and near the two longwall faces. The measurements show that the maximum principal stress is approximately parallel to the strike of the working face, moderate stress is parallel to the slope of the longwall face, and minimum principal

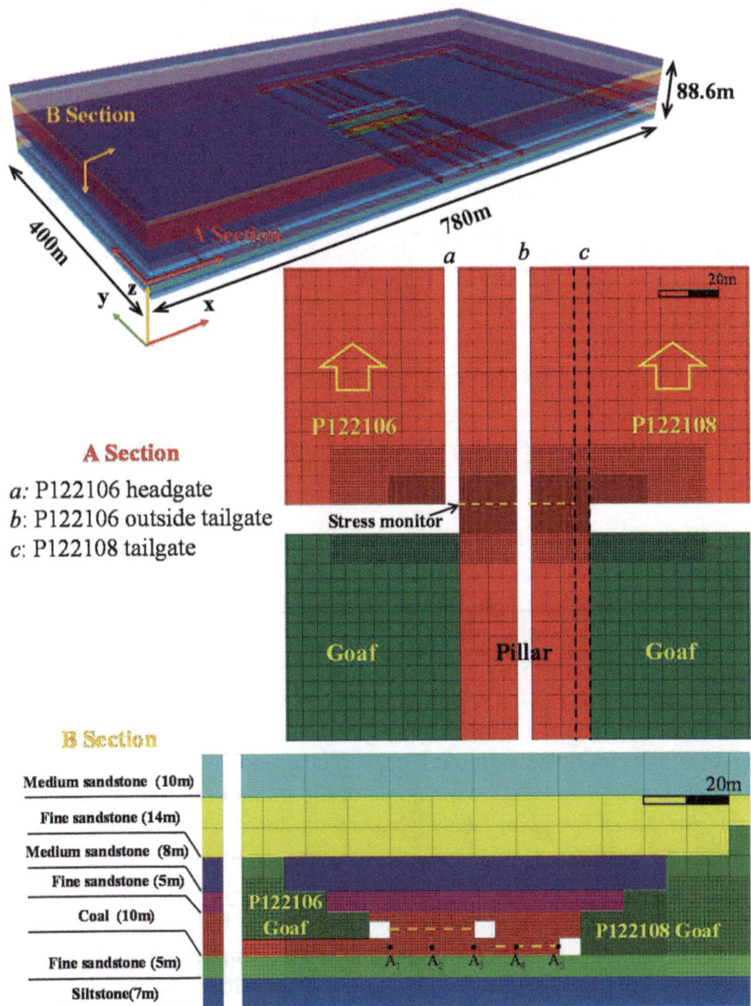


**Fig. 6.8** Numerical model and measuring point distribution of the longwall face

stress is the weight of the overburden with a direction perpendicular to the longwall face. The horizontal-to-vertical-stress ratios are $k_{max} = 1.5$ and $k_{min} = 1.2$. The bottom and side of the model are fixed with the horizontal displacement and vertical displacement, respectively. 5.95 MPa vertical stress is applied on top of the model to replace the load of the overlying strata.

In the model, the strain-softening failure criterion is used in all strata (Yang et al., 2011), and the physical and mechanical parameters are shown in Table 6.1 (the parameters in the table are the final parameters, which are obtained from the mining stress inversion below). Since the gradual compaction process of the goaf affects the stress distribution around the longwall face, the double-yield model is adopted for

**Table 6.1** Physical and mechanical parameters of rocks and coal

| Lithology | Bulk modulus (GPa) | Shear modulus (GPa) | Internal friction angle (°) | Tensile strength (MPa) | Cohesion (MPa) | Density (kg/m$^3$) |
|---|---|---|---|---|---|---|
| Medium sandstone | 5.1 | 2.7 | 27(20) | 1.8 | 6.5(0.65) | 2500 |
| Fine sandstone | 5.4 | 3.6 | 23(18) | 2.4 | 5.8(0.58) | 2350 |
| 2–2#Coal | 5.5 | 4.2 | 35(25) | 1.02 | 4.7(0.47) | 1400 |
| Siltstone | 8.3 | 6.2 | 32(23) | 3.1 | 7.5(0.75) | 2500 |

**Table 6.2** Cap pressure table of the double-yield goaf elements

| Strain | 0.00 | 0.02 | 0.05 | 0.07 | 0.1 | 0.12 | 0.15 | 0.17 | 0.20 |
|---|---|---|---|---|---|---|---|---|---|
| Pressure/MPa | 0.00 | 0.10 | 0.30 | 0.60 | 1.25 | 2.25 | 5.00 | 10.00 | 20.00 |

the goaf after excavation, and the cap pressure is shown in Table 6.2. After the initial stress balance, the tailgate, headgate, P122106 outside tailgate, and P122108 tailgate are excavated at the same time. The P122106 longwall face is excavated to 400 m at 10 m per cycle along the *Y* direction. After P122106 face excavation, the water is stored in the goaf with a water level of 6 m (Fig. 6.5), and then the P122108 longwall face is excavated to 400 m at 10 m per cycle along the *Y* direction. According to the calculation methods of Alber et al. (2009) and Qian et al. (2010), the height of the caving zone in the P122106 goaf and P122108 goaf is 17.6 m and 26.2 m, respectively, and the caving angle of the immediate roof is 30–45°.

In the process of longwall face advancement, due to the influence of front abutment pressure and lateral abutment pressure, the coal pillar will generate plastic deformation. In previous numerical simulation analyses (Zhang et al., 2017b), it was difficult to describe the development area of the plastic zone in the coal pillar with only one section, and there was a lack of a quantitative standard. To reasonably and quantitatively describe the damage ratio of coal pillars caused by longwall face mining, this chapter proposes the concept of a coal-pillar damage ratio, which refers to the ratio of the total volume of plastic zones in the monitoring area (or the total coal pillar) to the monitoring area volume during the longwall face advance, as shown in Fig. 6.9.

$$D = \frac{\sum_{i=1}^{n} V_i}{V_o} \times 100\% \tag{6.1}$$

where $D$ is the damage ratio; $V_i$ represents the volume of a plastic zone; $V_o$ is the volume of a monitoring area, $V_o = L \times S \times H$; $H$ is the height of a monitoring area, m; $L$ is the length of a monitoring area, m; and $S$ is the width of a monitoring area, m.

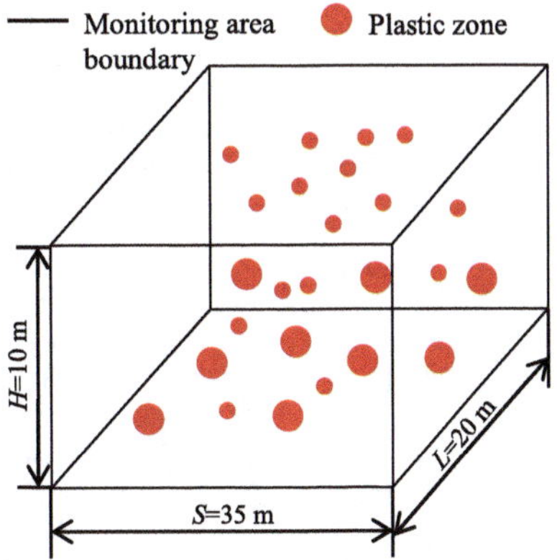


**Fig. 6.9** Schematic diagram of the calculation model of the coal-pillar damage ratio

To ensure the safety of the gate road system after water storage in the goaf, the width of the elastic core of the coal pillar should be no less than two times the mining height (Yao et al., 2019b). Therefore, the critical damage ratio of coal pillars $D_L$ is:

$$D_L = \frac{S - 2M}{S} \times 100\% \tag{6.2}$$

where $M$ is the mining height, m.

### 6.3.2 *Mining Stress Inversion of Coal Pillars*

To accurately obtain the parameters of each rock stratum in the numerical simulation, this chapter uses the measured results of the coal-pillar mining stress to invert the numerical simulation. The measured stress and numerical simulation stress distribution inside the coal pillar during the P122106 longwall face retreat are shown in Fig. 6.10. Comparing the numerical simulation results with field measurement results, it can be seen that the stress magnitude and evolutionary law are the same. Therefore, the inversion of stratum parameters can be used for further stability analysis of coal pillars.

### 6.3.3 *Stability Analysis of Coal-Pillar Mining-Water Invasion Weakening*

During the P122106 longwall face advancement to the monitoring area, the vertical stress of the pillar rib and pillar center increases, and the deformation of the P122108 tailgate remains stable, as shown in Fig. 6.11. As the P122106 longwall face mines

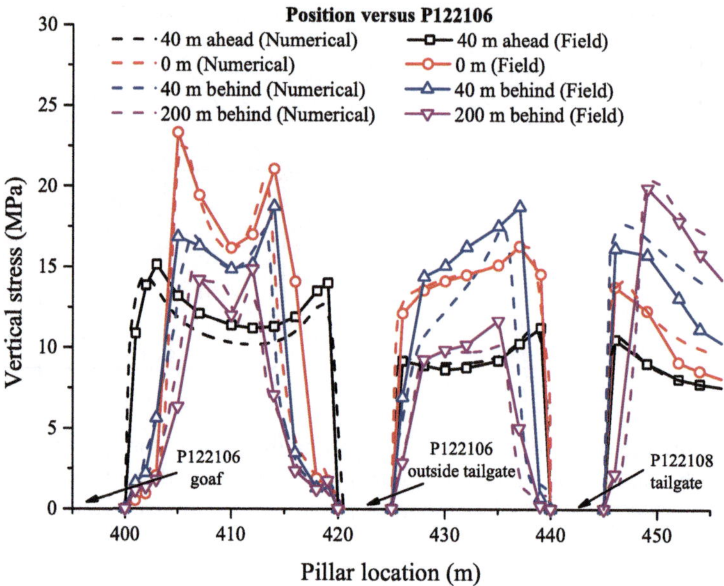


**Fig. 6.10** Stress distribution and its evolution in a coal pillar obtained by field measurements and numerical simulation

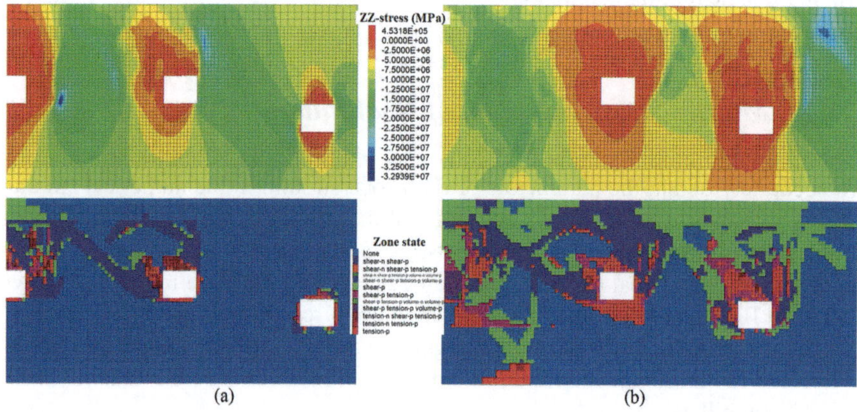


**Fig. 6.11** Comparison of induced stress and failure zone distribution. (**a**) P122106 advancement to 200 m. (**b**) P122106 advancement to 400 m

to the monitoring area, the vertical stress of the pillar rib decreases significantly, while the vertical stress of the pillar center continues to increase, and the deformation of the coal-pillar rib of the P122108 tailgate begins to increase. As the longwall face continues to advance, the coal body at the pillar rib has been in a state of residual stress, and its vertical stress remains stable. The vertical stress at the center pillar reaches the peak value 20 m ahead of the P122106 face and then decreases gradually, and the deformation of the P122108 tailgate increases gradually in an "S" shape. During P122106 face retreat, the stress concentration area formed by mining disturbance is transferred from the left-side pillar of the P122106 outside tailgate to the right-side pillar and then finally is transferred to the solid coal slope of the P122108 longwall face, as shown in Fig. 6.11.

After mining of the P122106 longwall face, water began to be stored in the goaf with a water level of 6 m. As the P122108 longwall face closes on the monitoring area, the vertical stress in the coal pillar remains stable (residual stress state), and the vertical stress and pore pressure of the pillar center show an increasing trend. When the monitoring area is 50 m ahead of the P122108 longwall face, the goaf water has infiltrated into the measuring point at the pillar center, and the deformation of the P122108 tailgate increases linearly. As the longwall face continues to advance to the monitoring area, the bearing capacity of the coal pillar is weakened, and the vertical stress in the coal pillar is reduced by 3.7 MPa compared with the vertical stress at the same position of the P122106 face. At this time, the deformation of the tailgate is approximately 1.4 m accompanied by the water seepage phenomenon in the pillar rib. As shown in Fig. 6.12a, the numerical simulation deformation of the tailgate is consistent with field measurement results.

According to the numerical simulation method of coal-pillar stability analysis under mining-water invasion, 40 cycles are carried out for each 10 m of excavation of the P122108 face (50 steps of mechanical calculations and 50 steps of fluid calculations in 1 cycle). The changes in the water invasion weakening parameters, pore pressure, stress, damage ratio, and plastic zones of coal pillars under different cycles are monitored and analyzed. Figure 6.13 shows the pore pressure and mechanical parameters weakening at different positions in the coal pillar after different iterations. The specific layout of coal-pillar monitoring point is shown in Fig. 6.8.

Figure 6.13 shows that there is a certain difference in the pore pressure of each monitoring point in the coal pillar, which causes the weakening of the mechanical parameters to be different. Measuring points $A_1$ and $A_5$ were yielded in the process of P122106 face mining. Because the plastic zone in the coal pillar is not connected, only the permeability of measuring point $A_1$ increases sharply. With increasing water pressure, the water content at measuring point $A_1$ greatly increases, which leads to a decrease in the internal friction angle and cohesion, resulting in the further development of the plastic zone in the interior of the coal pillar. During P122108 face mining, the stress in the pillar ribs is obviously less than that in the pillar center (Fig. 6.12b), and measuring points $A_1$ and $A_5$ have higher permeability. In addition, $A_1$ is closer to the goaf, and the water pressure in the goaf preferentially flows into $A_1$. This leads to a prior increase in water pressure at $A_1$, and the final stable value is higher. When fluid-solid coupling calculations are iterated 960 times (the working face is 40 m ahead of the monitoring point), the friction angle and cohesion at each measuring point tend to be stable gradually. To further analyze the distribution of water pressure and the plastic zone in the coal pillar during different iterations, this chapter selects the distribution of plastic zones and pore pressure after the 40th, 400th, 960th, and 1200th iterations, as shown in Fig. 6.14.

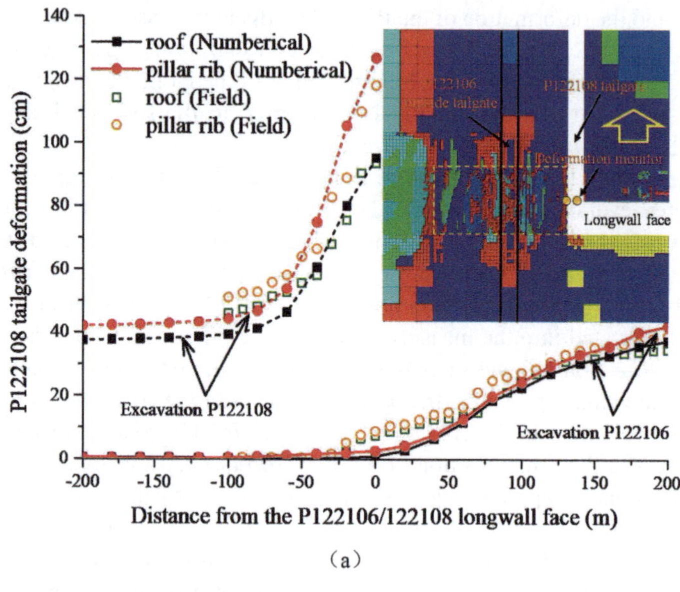


(a)

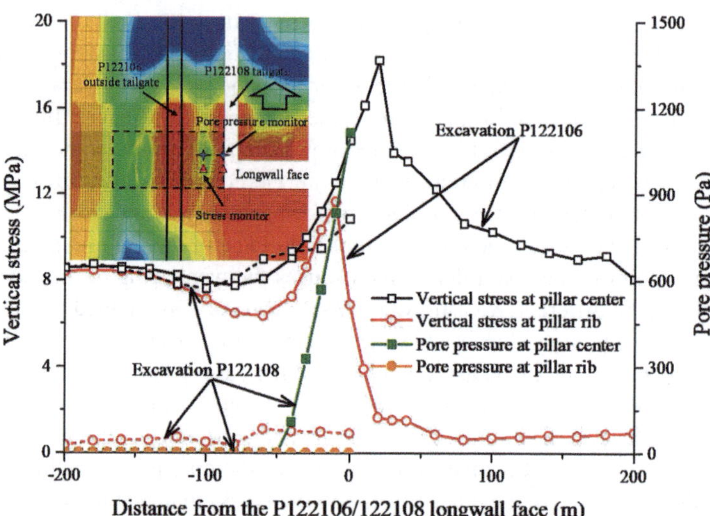


**Fig. 6.12** Bearing characteristics of the coal-pillar face during P122106/P122108 face advancement. (**a**) P122108 tailgate deformation. (**b**) Distribution of coal-pillar stress and pore pressure

As shown in Fig. 6.14, when the number of iterations is 40, P122106 goaf water flows into the surrounding yield coal and rock mass. Measuring points $A_2$ and $A_4$ are elastic zones, $A_1$, $A_3$, and $A_4$ are plastic zones, and the water flow from plastic zones to elastic zones is very slow. When the number of iterations is 400, the goaf water has penetrated the coal-pillar bottom of the P122106 outside tailgate and plastic zones in the coal pillar develop at 28 m. The measuring point $A_2$ changes from an

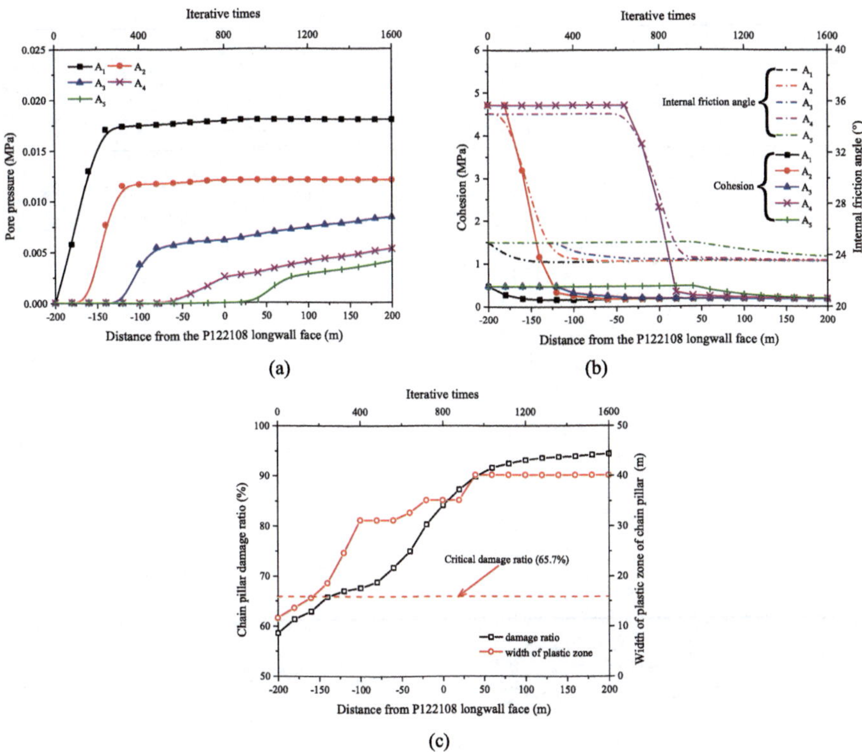


**Fig. 6.13** Pore pressure, damage degree, and strength weakening in the coal pillar under different iterations (longwall face advancement distance). (**a**) pore pressure, (**b**) mechanical parameters, and (**c**) damage ratio of the monitoring area.

elastic zone to a plastic zone, and the pore pressure increases in an "S" shape. When the number of iterations is 960, measuring point $A_2$ changes from the elastic zone to the plastic zone, and the plastic zone at the bottom of the coal pillar is connected. When the number of iterations increases to 1200, the range of the plastic zone in the coal pillar is still slowly increasing, and the vertical stress decreases gradually, resulting in a slow increase in pore pressure at measuring points $A_3$, $A_4$, and $A_5$.

As shown in Fig. 6.13c, the development width of the plastic zone and the pillar damage ratio increase gradually with an increasing number of iterations. When the plastic zones in the coal pillar are connected, the damage variable is still increasing, while the development width of the plastic zone in the coal pillar remains constant. At this time, the damage ratio can display the characteristics of asymptotic damage in the coal pillar during longwall face mining. When the number of iterations increases to 800, the coal-pillar damage ratio in the monitoring area is 84.1%, which exceeds the critical damage ratio of coal-pillar instability (65.7%). Therefore, based on the condition that the water level in the P122106 goaf is 6 m, the 35 m wide coal pillar cannot block the goaf water flowing into the P122108 longwall face and cannot maintain P122108 tailgate stability.

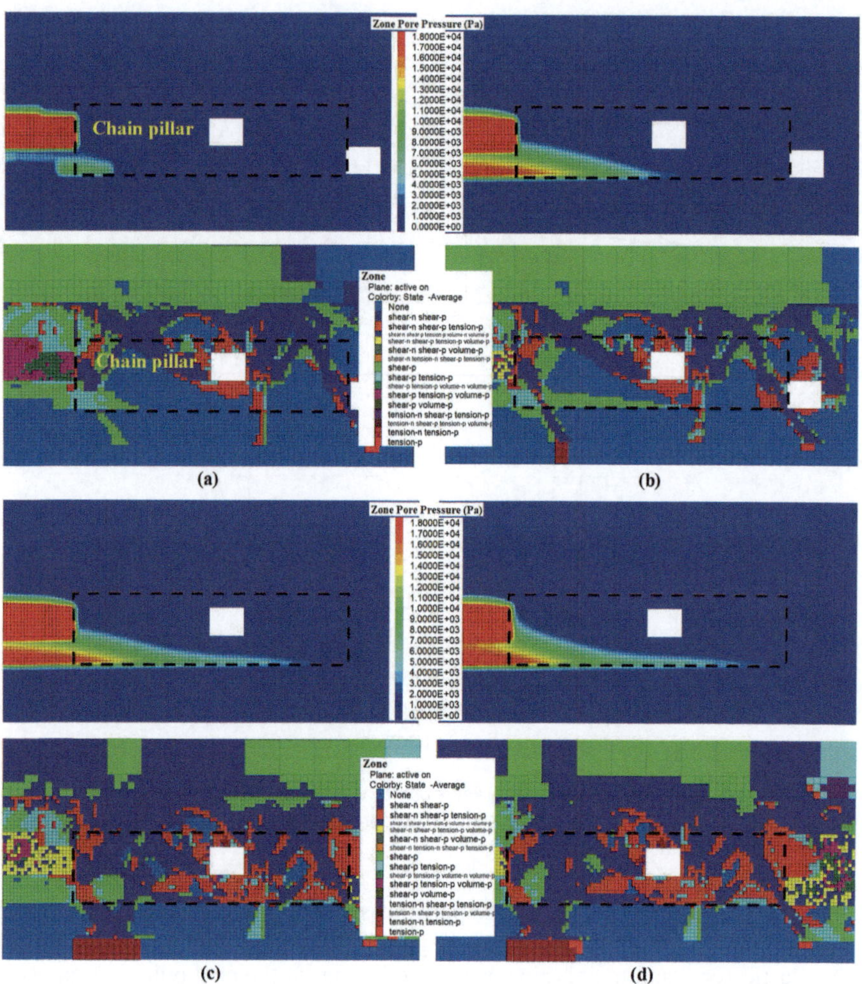


**Fig. 6.14** Distribution of pore pressure and the plastic zone in coal pillar under different cycles: (**a**) excavation 10 m (40 iterations), (**b**) excavation 100 m (400 iterations), (**c**) excavation 240 m (960 iterations), and (**d**) excavation 300 m (1200 iterations)

According to the above simulation, the coupling of stress and water results in the instability of the coal pillar. Therefore, two main methods can be used to maintain the stability of the coal pillar. The first is to set up a wider coal pillar to bear the high abutment stress through the high stress in the coal pillar to reduce the permeability of the coal pillar, blocking the seepage path of water. The other method is to reduce the height of the water level in the goaf, which results in a lower water pressure on the coal pillar and slows down the water-immersion weakening. At the same time, strengthening support is used to reduce the large deformation of the gate road.

## 6.4 Stability Control Measures for Coal Pillars

### 6.4.1 *Suitable Pillar Size*

In the longwall-mining process, different widths of a coal pillar can effectively change the distribution characteristics of abutment stress in the coal pillar. Coal permeability decreases slowly with increasing stress before coal pillar failure and increases sharply when plastic failure of the coal pillar occurs (Meng et al., 2016). When a coal pillar sufficiently bears mining stress, the high-stress region within the coal pillar can effectively block the seepage path of water. During P122108 longwall mining, the water level in the P122106 goaf is maintained at approximately 6 m (Fig. 6.5). Figure 6.15 shows the plastic zone and vertical stress distribution characteristics of 35 m, 45 m, and 55 m coal pillars at a water level of 6 m. Two stages in the longwall-mining process are selected, in which stage I is P122106 longwall face to 400 m, and stage II is P122108 longwall face to 200 m.

The specific analysis of a 35 m wide coal pillar is presented in Sect. 3.3. When a pillar size of 45 m was used, an elastic zone of 25 m is retained in the coal pillar, and the vertical stress at the pillar ribs is concentraated and less constrained at stage I, resulting in instability in the pillar ribs. The vertical stress in the coal pillar is mainly concentrated on the right side of the P122106 outside tailgate, and the damage ratio of the coal pillar is 44.9% (Fig. 6.15b). The P122108 face continues to advance to stage II; the plastic zone does not fully run through the coal pillar, and the damage ratio of the coal pillar is 65.3%. At this time, the pillar and gate road must bear the high stress in front of the longwall face at a distance of 40 m, while the peak abutment stress in the solid coal side in front of the longwall face is smaller than that in the 35 m width coal pillar.

When the coal pillar increases to 55 m, at stage I, the plastic zone in the coal pillar develops at 2.5 m, the damage ratio is 37.1%, and the stress distribution is similar to that of the 45 m wide coal pillar. When the P122108 face continues to advance to stage II, the coal pillar can fully bear the mining stress and softening effect of the goaf water, and the high stress is concentrated in the interior of the coal pillar, as shown in Fig. 6.14c. At this time, the damage ratio of the coal pillar is 47.4%, which is much less than its critical damage (78.2%).

The evolution of the stress and plastic zones causes an obvious difference in the deformation of the P122108 tailgate during P122106 and P122108 face advancement. The coal pillar with a width of 35 m shows characteristics of strain softening. At stage I, the 35 m coal pillar still maintained a certain strength, and the deformation of the P122108 tailgate was 35–40 cm. When the P122108 face advances, the coal pillar with a width of 35 m begins to soften. The deformation of the tailgate increases linearly (Fig. 6.12), indicating that the coal pillar can no longer fully bear the mining stress and the effect of water-immersion softening in the P122106 goaf. Therefore, a large part of the roof load is transferred to the solid coal side of the tailgate. At stage II, the 45 m and 55 m pillars can bear most of the roof load; part of the load is transferred to the solid coal side, and the deformation of the tailgate is

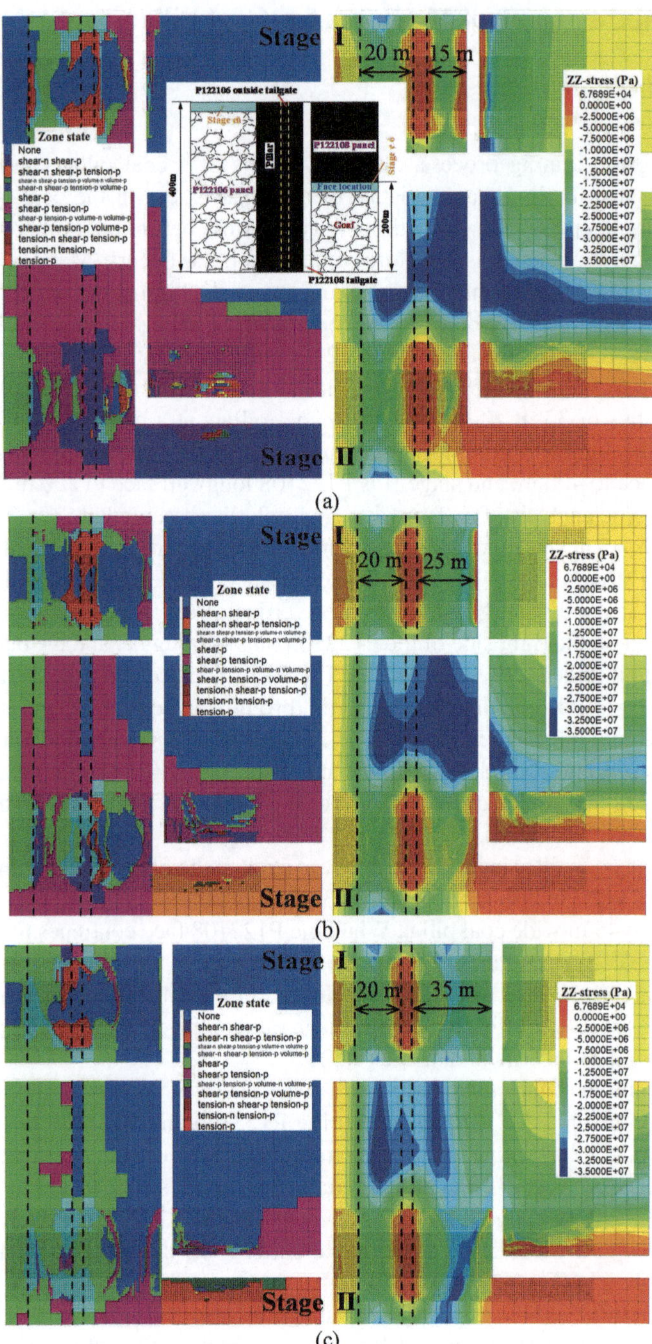


**Fig. 6.15** Plastic zone and stress distributions with pillar sizes of (**a**) 35 m, (**b**) 45 m, and (**c**) 55 m at stages I and II. (The definitions of Stages I and II are shown in Fig. 6.17a)

65–70 cm and 20–25 cm, respectively, which is 41.7% and 79.2% lower than that of the 35 m coal pillar.

To avoid P122106 goaf water flow into the P122108 longwall face, it is necessary to retain a certain bearing area in the coal pillar. The variation in the coal-pillar damage ratio with the advancement distance of the P122108 face is shown in Fig. 6.16b. According to formula (5), the critical damage in the 35 m, 45 m, and 55 m coal pillars can be obtained by 65.7%, 73.3%, and 78.2%, respectively. During P122106 face mining, all three sizes of coal pillars can maintain the stability of the tailgate, but the damage ratio of the 35 m coal pillar exceeds its critical damage ratio, resulting in the emergence of rib spalling and the inflow of goaf water into the P122108 face. The damage ratio with a 45 m wide coal pillar is greater than its critical damage ratio after the P122108 face advances to 320 m, which indicates that a 45 m wide coal pillar cannot prevent goaf water from flowing into the longwall face during P122108 face mining. Moreover, when P122108 advances to the monitoring area, the numerical results show that the pillar rib deformation and roof deformation of the P122108 tailgate are 66.2 cm and 56.1 cm, respectively, which indicates that it was noticeably difficult to control the deformation of the tailgate. When the coal-pillar size increases to 55 m, the coal-pillar damage ratio is less than its critical damage ratio. Therefore, the numerical simulation suggested that the coal-pillar size should be 55 m to ensure the stability of the coal pillar and tailgate during P122108 mining.

## 6.4.2 Suitable Height of the Goaf-Water Level

Through the above research, it is found that an increase in coal-pillar size is beneficial to the stability of the coal pillar and gate road, but with an increase in mining height, reducing the coal-pillar size can achieve obvious economic benefits. For example, the coal-pillar size of the P122108 face decreased by 1 m, and the coal

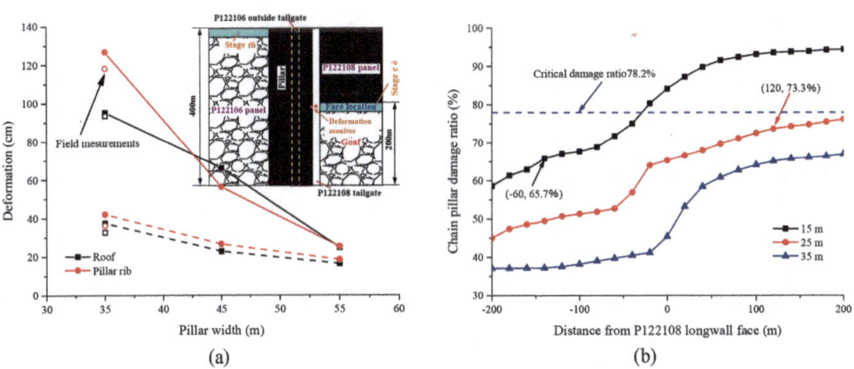


**Fig. 6.16** The P122108 headgate deformation and pillar damage ratio of different pillar sizes from numerical simulations. (**a**) P122108 headgate deformation. (**b**) Damage ratio of the coal pillar

output increased by 3846 t. The method of reducing the goaf water level height can cause the coal pillar to bear the appropriate water pressure and can slow down the weakening effect from the water on the coal pillar. Therefore, this method is an effective measure to reduce the coal-pillar width and maintain gate road stability.

The stability of the coal pillar and gate road is numerically simulated by controlling the water level in the goaf. The height of the caving zone of the P122106 face is 17.6 m. Therefore, the maximum height of the water level in the goaf is assumed to be 18 m, and the water level in the goaf is selected as 0 m, 6 m, 12 m, and 18 m during the numerical simulation, as shown in Fig. 6.19. In the other subfigures, the results are shown from the same cross section when the P122108 face is mined to 200 m.

Figure 6.17a shows that reducing the height of the goaf water level decreases the pore pressure in the coal pillar, which could increase the bearing capacity of the coal pillar and subsequently relieve the deformation of the tailgate. When the size of the coal pillar is 35 m, the plastic zone has run through the upper part of the coal pillar, as shown in Fig. 6.17a. When the P122106 goaf water level is 0 m, the coal mass in the middle of the coal pillar and the bottom of the outside tailgate remain elastic zones, and the stress concentration in the coal pillar is saddle-shaped. When the goaf water level increases, the plastic zone at the bottom of the coal pillar gradually expands to the P122108 longwall face during P122108 face mining. When the water level increases to 12 m and 18 m, the pore pressure at the coal-pillar rib of the P122108 tailgate is greater than 0 MPa, indicating that the yield zone completely runs through the coal pillar and that the goaf water will seep into the P122108 tailgate through the plastic zone of the pillar. Compared with the P122106 goaf water level at 0 m, when the water level is 18 m, the maximum deformation of the coal-pillar rib and roof in the P122108 tailgate increases to 154.0 cm and 117.7 cm from 82.1 cm and 80.8 cm, respectively, as shown in Table 6.3. When the goaf water level increases, the small difference in mining stress in the coal pillar indicates that the 35 m coal pillar is damaged under different water levels.

When the size of the coal pillar is 45 m, the plastic zone still runs through the upper part of the coal pillar, but the stress in the pillar presents different profiles compared with the pillar width of 35 m, as shown in Fig. 6.17b. When the goaf water level is 0 m and 6 m, the stress distribution in the coal pillar is saddle-shaped, and the stress concentration coefficient gradually increases with increasing goaf water level. When the goaf water level increases to 12 m and 18 m, the stress distribution in the coal pillar changes from a saddle shape to a unimodal distribution, and the stress concentration coefficient decreases gradually with increasing goaf water level. As the P122108 face is advanced to 200 m, the plastic zone at the bottom of the coal pillar does not run through under the four water levels, and the water in the goaf does not flow into the working face. Compared with the goaf water level at 18 m, when the water level is 6 m, the maximum deformation of the coal-pillar rib and roof of the P122108 tailgate decreases to 66.1 cm and 56.2 cm from 85.3 cm and 79.5 cm, respectively, as shown in Table 6.3, which indicates that it was noticeably difficult to maintain the stability of the tailgate during P122108 face mining.

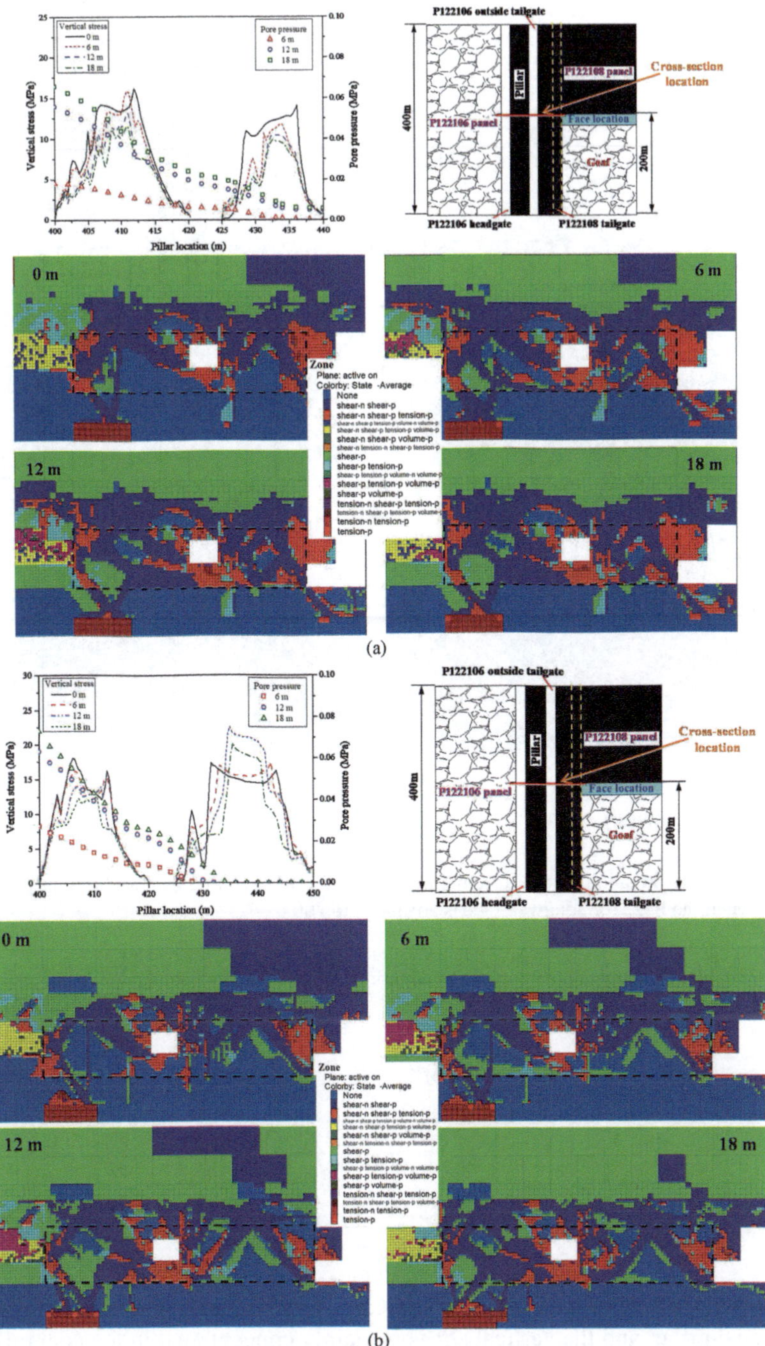


**Fig. 6.17** Distribution of mining-induced stress, plastic zone, and pore pressure with different coal-pillar sizes when the goaf water level is 0 m, 6 m, 12 m, and 18 m: (**a**) 35 m, (**b**) 45 m, and (**c**) 55 m

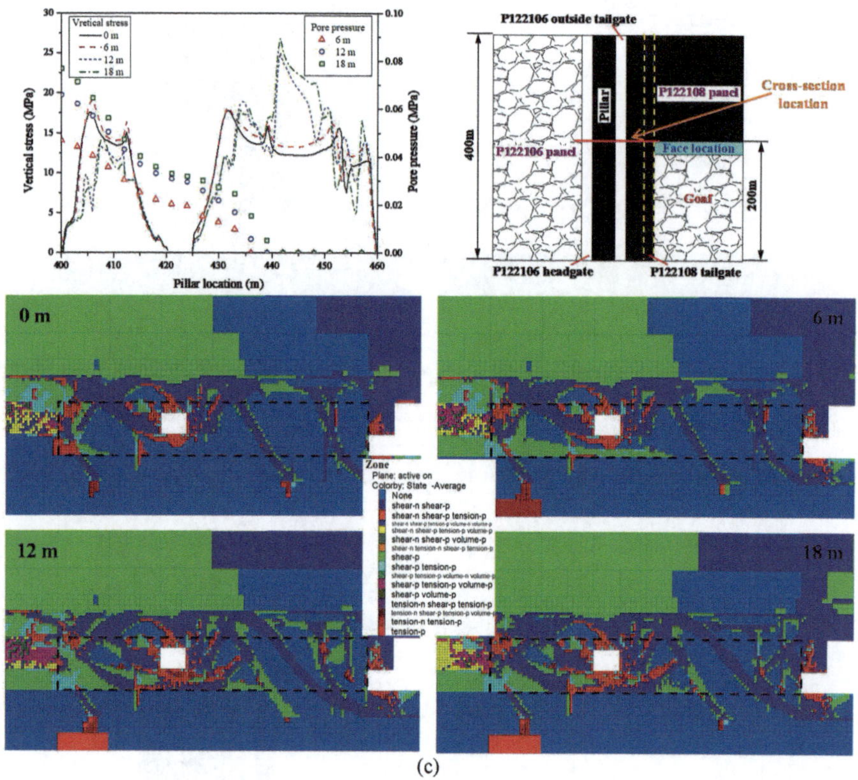


**Fig. 6.17** (continued)

**Table 6.3** Deformation of the surrounding rock of the P122108 tailgate at different goaf water level when the P122108 longwall face is advanced to 200 m

| P122106 goaf water level (m) | Deformation (cm) | Pillar sizes (m) | | |
|---|---|---|---|---|
| | | 35 | 45 | 55 |
| 0 | Roof | 80.8 | 54.2 | 21.1 |
| | Pillar rib | 82.1 | 50.3 | 21.2 |
| 6 | Roof | 95.2 | 56.2 | 24.7 |
| | Pillar rib | 126.5 | 66.1 | 25.1 |
| 12 | Roof | 103.5 | 67.2 | 37.2 |
| | Pillar rib | 141.2 | 72.7 | 36.2 |
| 18 | Roof | 117.7 | 79.5 | 42.0 |
| | Pillar rib | 154.0 | 85.3 | 44.5 |

When the size of the coal pillar increases to 55 m, an elastic core is formed in the coal pillar, and the vertical stress is mainly concentrated in the coal pillar, as shown in Fig. 6.19c. When the goaf water level is 0 m and 6 m, the vertical stress distribution state is similar to that of a 45 m wide coal pillar. When the goaf water level increases to 12 m and 18 m, the stress distribution in the coal pillar also changes from a saddle shape to a unimodal state, but the stress concentration coefficient gradually increases with the increase in the goaf water level, and a high stress of 26.8 MPa is concentrated in the coal pillar. The maximum deformation of the coal-pillar rib and roof of the tailgate under the four water-level conditions is less than 45 cm, as shown in Table 6.3. With the increase in the goaf water level, the pore pressure is distributed on the left side of the peak stress in the coal pillar, which indicates that the pillar was sufficiently strong to support the ever-increasing roof loading, and the pillar remained in a stable state during the entire process.

Figure 6.18 shows the variation in the coal-pillar damage ratio with the goaf water level height of P122106 when P122108 advances to 400 m under different coal-pillar sizes. The increase in the water level in the goaf leads to an obvious increase in the internal damage of the coal pillar. When the width of the coal pillar is 35 m, the pillar damage ratio under different water levels exceeds its critical damage ratio (65.7%) indicating that plastic zones have been completely run through the coal pillar, forming a water inrush channel. When the width of the coal pillar increases to 45 m and the goaf water level is lower than 4.1 m, the damage ratio of the coal pillar is less than the critical damage (73.3%) which shows that controlling the goaf water level below 4 m can ensure the stability of the coal pillar and avoid goaf water flowing into the tailgate during the mining of the P122108 face. For the 55 m wide coal pillar, the pillar damage ratio under different water levels is less than its critical damage (78.2%), which can ensure the stability of the coal pillar and tailgate in the mining process of the longwall face.

In conclusion, the lowering of the goaf water level will slow down the development of plastic zones in the coal pillar and will enhance the bearing capacity of the coal pillar, therefore causing smaller gate road deformations. When the width of the coal pillar is 35 m, controlling the water level of the goaf cannot meet the requirement of water isolation for the coal pillar. When the width of the coal pillar is 45 m, controlling the goaf water level below 4 m can achieve the requirement of water isolation for the coal pillar in the mining process of the P122108 face, but a significant increase in headgate deformation could be inevitable during the mining of the P14102 face. When the width of the coal pillar increases to 55 m, the coal pillar can satisfy the stability of the coal pillar and tailgate under different goaf water level. In this case, raising the goaf water level height could increase the stress concentration in the coal pillar, effectively reducing the permeability in the coal pillar and preventing the seepage path of water, and the increase in the water level has no obvious effect on tailgate deformation. Lowering the goaf water level can significantly reduce tailgate deformation with a pillar size of 45 m. Therefore, the 45 m coal pillar can maintain the stability of the tailgate and coal pillar by also controlling the P122106 goaf water level within 4 m and adopting reinforced support.

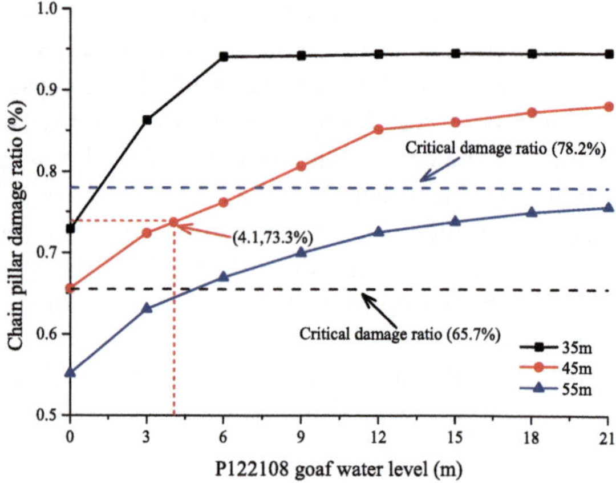


**Fig. 6.18** The change in the coal-pillar damage ratio with water level height in the P122106 goaf when the P122108 face is advanced to 400 m under different coal-pillar sizes

### 6.4.3 Field Practice

When P122108 mining reaches 1500 m, serious deformation occurs in the tailgate, and water inrush occurs in the bolt hole on the side of the coal pillar (Fig. 6.7b). If the 45 m coal pillar is left to re-excavate the P122108 tailgate at this time, it will cause a reduction in the longwall face productivity and the high cost of moving the longwall face. Therefore, P122106 goaf water is drained to ensure safety in the production of the longwall face, and reinforced support is used to reduce the deformation of the tailgate. The headgate, outside tailgate, and tailgate of the P122106 face are sealed with reinforced concrete, and pumps are installed to drain the P122106 goaf water (Fig. 6.19a).

To reduce the deformation of the tailgate, the support is strengthened according to the structure of the roadway, which is easy to destabilize. According to the above analysis, the weakest structures of the tailgate are the coal pillar and the roof. Three additional anchor cables are added to the tailgate roof, two additional anchor cables are added to the middle and upper parts of the coal-pillar rib, and a steel belt is used to support the coal-pillar rib. The specific support scheme is shown in Fig. 6.20.

After P122106 goaf water is discharged, reinforced support is applied to the roadway in front of the working face. Figure 6.21 shows the variation in the deformation of the P122108 tailgate with the advancing distance of the P122108 face. When the P122108 face is advanced to the monitoring area, the deformation of the coal-pillar rib and roof of the tailgate is 118.0 cm and 93.5 cm, respectively (Fig. 6.6b). After draining goaf water and strengthening the tailgate support, the deformation rate of the tailgate is significantly reduced, and the floor drum is effectively controlled during P122108 mining (Fig. 6.19). As shown in Fig. 6.21, when the P122108 face advances to the monitoring area, the deformation of the coal-pillar rib and roof of the tailgate is 71.3 cm and 64.7 cm, respectively. Compared with the deformation of the tailgate with a water level of 6 m in the goaf, the deformation of

**Fig. 6.19** Treatment of P122106 goaf and tailgate failure after reinforced support. (**a**) Drainage device in the goaf. (**b**) Coal-pillar rib deformation. (**c**) Floor cracking and floor heave. (**d**) Solid coal deformation

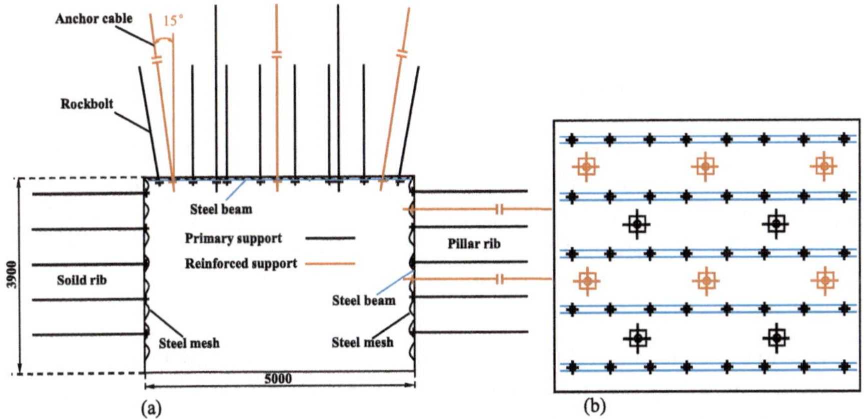


**Fig. 6.20** Strengthened support scheme. (**a**) Front view of support. (**b**) Top view of support

the coal-pillar rib and roof decreases by 39.6% and 30.8%, respectively. The results show that when the width of the coal pillar cannot meet the gate road stability, reducing the height of the goaf water level and strengthening support could effectively reduce the deformation and maintain gate road stability.

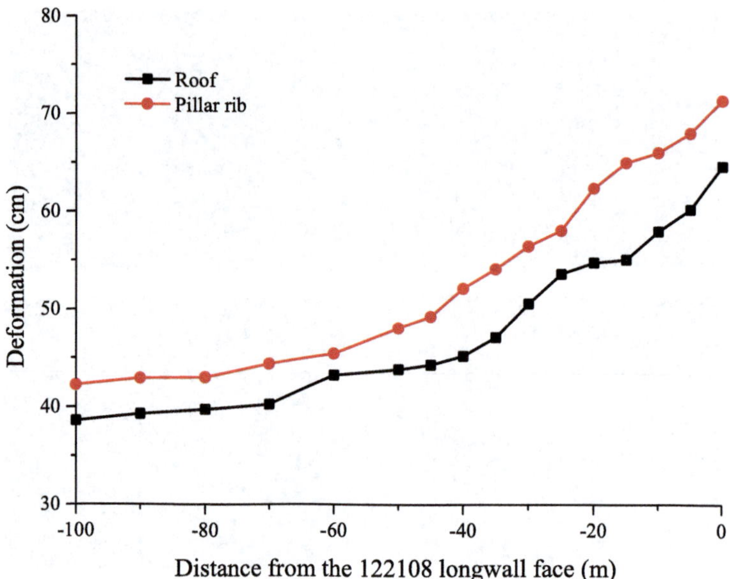


**Fig. 6.21** Variation law of the P122108 tailgate deformation during P122108 face mining (goaf water level at 0 m and reinforcement support)

## 6.5 Conclusions

In this chapter, according to the characteristics of water invasion softening of coal masses, a fluid-solid coupling numerical simulation method is proposed. Combined with field measurement results, distribution of mining stress, displacement of the coal pillar and roof, and the water-seepage situation of the P122108 longwall face under the mining-water invasion coupling effect are analyzed and evaluated. On this basis, the index of the coal-pillar damage ratio is defined, and the critical damage ratio of coal-pillar instability is given, which is helpful for quantitative analysis of the coal-pillar damage ratio and judgment of coal-pillar instability. The results show that:

1. Due to the weakening effect of P121106 goaf water, a coal pillar with a width of 35 m cannot fully bear the roof load, and an obvious phenomenon of water seepage occurs in the coal-pillar rib of the P122108 tailgate. After coal-pillar failure, the concentrated stress is transferred to the solid side of the P122108 face, which further results in a substantial increase in roadway deformation, indicating that the existing support conditions and width of the coal pillar are not sufficient to isolate the goaf reservoir, and there is a danger of a water inrush in the longwall face.
2. When a longwall face with different coal-pillar widths advances to the monitoring area, the strength of a 35 m coal pillar decreases rapidly when it reaches its ultimate strength, which is the main reason for the deformation and instability of

the tailgate and the phenomenon of water seepage in the coal-pillar rib. When the width of the coal pillar increases to 45 m, there could be water seepage in the coal-pillar rib of the tailgate after the P122108 face advances to 320 m. Moreover, when the P122108 face is close to the monitoring area, the P122108 tailgate deformation still increases rapidly. At this time, if the P122106 goaf water level is decreased, goaf water can be prevented from flowing into the P122108 face, and the tailgate deformation can be effectively reduced. When coal-pillar size is larger than 55 m, the coal pillar and tailgate can remain stable because there is still a certain width of elastic zones in the coal pillar to block goaf water.

3. The influence of different goaf water level heights on the stability of coal pillars is studied. This shows that the pillar damage ratio can be effectively reduced by controlling the water level of the P122106 goaf, and the isolation size of the coal pillar can be decreased by reducing the goaf water level. This study provides a reasonable water level height corresponding to coal pillars of different sizes and finally determines that a coal pillar with a size of 45 m can maintain coal-pillar stability and can reduce the large deformation of the tailgate by controlling the P122106 goaf water level within 4 m and adopting reinforced support to the tailgate.
4. Considering the field conditions and numerical simulation results, this chapter proposes measures to drain the water in the reservoir of the P122106 goaf to ensure the safe production of the P122108 longwall face and to strengthen the supports to reduce damage and deformation of the tailgate. After using the abovementioned methods in the study mine, deformation of the coal-pillar rib and roof of the tailgate is reduced by 39.6% and 30.8%, respectively, ensuring the safe production of the longwall face.

# Chapter 7
# Progressive Failure of Roadways in a Fault Structure Area with Water-Rich Roofs

## 7.1 Introduction

China's proven coal reserves account for approximately 13.2% of the world's proven coal reserves, and coal mining has become the main way for China to obtain energy. With the continuous investment in coal mining safety, the accident rate and the number of deaths caused by coal mining are decreasing year by year (Zhang et al., 2021a). However, the geological conditions of the coalfield in China are complex. When the fault structure is connected to the aquifer, it will cause serious water gushing during the roadway excavation process (Ceccato et al., 2021; Salomon et al., 2021). Although it will not cause a water inrush accident, roadways have been in a state of water gushing and drenching for a long time, which will weaken the roadway surrounding rock (RSR). In addition, the surrounding rocks of the fault zone are seriously broken, which further increases the difficulty of supporting the roadway with a water-rich fault structure (Wang et al., 2020b). Therefore, it is necessary to analyze the instability mechanism of roadways in the region of water-filled fault structures, so as to put forward control measures.

A large number of laboratory tests show that the mechanical properties of rocks are negatively correlated with moisture content, especially mudstone and sandy mudstone with weak interlayers (Yao et al., 2019a, 2019b; Xiao et al., 2020; Ai et al., 2021). The water invasion weakening in rock masses is the result of the comprehensive action of physics (lubrication, scour transport, and disintegration) and chemistry (combined water strengthening, ion-exchange and oxidation-reduction, etc.). Groundwater causes internal damage to rock masses by changing the mineral contents and intergranular structure of the rock (Maruvanchery & Kim, 2019), thereby decreasing the macroscopic mechanical characteristics of rock masses (Fujii et al., 2020; Han et al., 2022a, 2022b). Therefore, the weakening effect of water is the main factor affecting the stability of the RSR, especially the RSR in the geological structure (Salmi et al., 2017).


C. Zhang, *Water Rock Interaction in Underground Coal Mining*,
https://doi.org/10.1007/978-981-95-9957-8_7

In recent decades, different scholars have studied the mechanism of fault activation induced by mining and the effect of advancing distance on fault water inrush through simplified mechanical models (Hu et al., 2019; Bu & Xu, 2020), which provides some guidance for predicting the critical water-resisting capacity of faults after longwall mining (Song & Liang, 2021). However, the stability of the RSR is affected by mining stress, RSR support, mining history, aquifer water pressure, and weakening of coal and rock masses (Bai et al., 2017; Zhang et al., 2019i; Showkati et al., 2021), there are significant limitations to research the complex progressive failure using a theoretical method. Therefore, many kinds of research used numerical methods and field measurements to study the mechanical response and progressive failure characteristics of surrounding rocks after roadway excavation, and abundant research achievements have been obtained (Li & Wu, 2019; Wang et al., 2019; Bai & Tu, 2020). However, most of the existing research results analyze the stability of RSR by studying fault activation and rarely involve the coupling effect of water invasion and mining. Bai et al. (2016) proposed a time-varying strength weakening model and applied it in the numerical model. However, the movement of water in RSR is not considered, and the assumption that the final strength linearly decreases to zero is not completely consistent with the experimental results. Sun et al. (2019) used a Comsol simulation to analyze the formation and evolution of water inrush passage from the floor of a deep fault mine. However, due to the lack of consideration of weakening of the mechanical and seepage parameters, the simulation results have errors.

In this chapter, taking the Beixinyao Coal Mine as the research object, the Universal Distinct Element Code (UDEC) Voronoi numerical model is adopted to study the progressive failure process of RSR when a roadway passes through the fault under the condition of water-rich roofs. Based on the quantitative relationship of water immersion weakening of rock samples in laboratory tests, a fluid-solid coupling simulation method and a damage quantitative index suitable for the engineering scale are put forward. Therefore, the progressive failure characteristics of the RSR under the roof aquifer are simulated. Finally, according to the progressive failure characteristics of RSR from water-rich roofs, a control strategy of dewatering the roof aquifer near the fault and strengthening the roadway support is proposed, which ensures the safe production of longwall face.

## 7.2 Field Measurement of Surrounding Rock Failure of Roadway

### 7.2.1 *Engineering Geological Condition*

The Beixinyao Coal Mine is situated in northern Shanxi Province, as shown in Fig. 7.1a. At present, the #2 coal seam is mainly mined. The coal seam is near the east-west direction with a dip angle of 6–23°, and a mean thickness of 6.2 m.

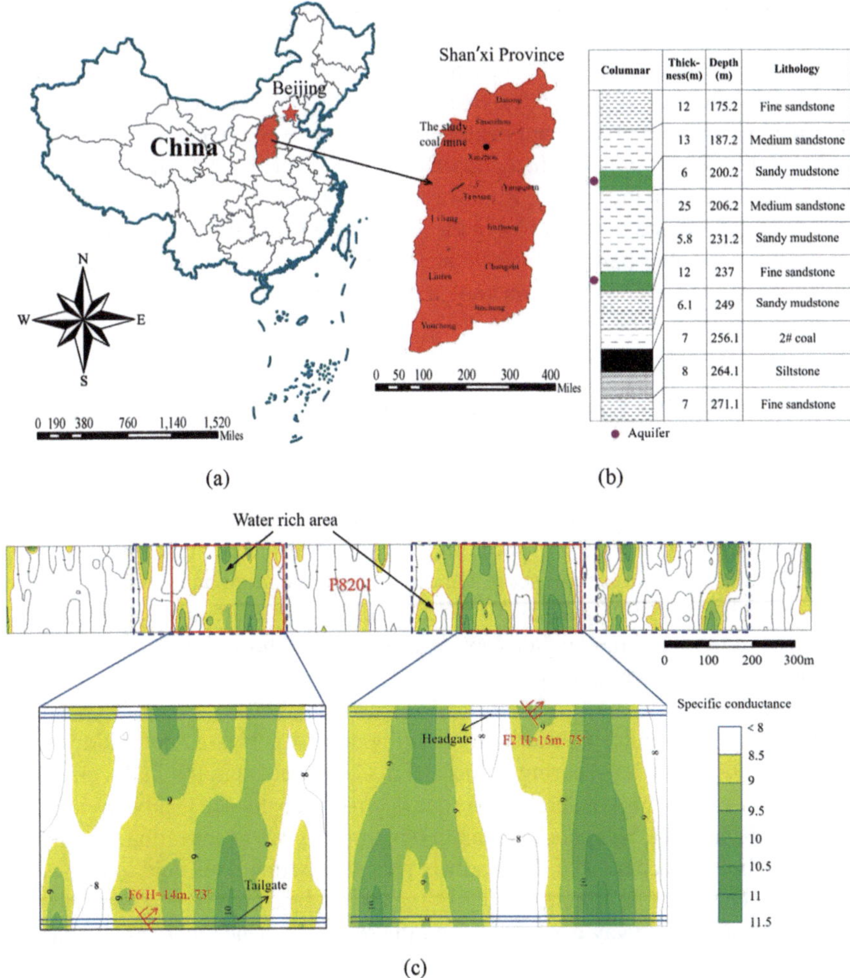


**Fig. 7.1** (**a**) Coal mine location, (**b**) stratigraphic column, and (**c**) specific conductance distribution above P8201

Aquifers above the #2 seam are Permian sandstone aquifers and Ordovician limestone aquifers, which are moderate to strongly water-rich. The average thickness of the aquifers is 6 m, and they are distributed mainly in sandy mudstone and belong to weakly fractured aquifers. The stratigraphic column is shown in Fig. 7.1b. Among these parameters, the measured water pressure of the Ordovician limestone aquifer is 1.8 MPa. The distribution of the roof aquifer within 30 m of #2 seam is detected by the audio frequency electric penetration method, as shown in Fig. 7.1c. When the conductivity of rock strata is greater than the threshold value (8.5), this area is abnormal, and the roof strata with higher conductivity have stronger moisture content.

P8201 is the first longwall face in the Beixinyao Coal Mine. The P8201 adopts top coal caving, and the layout of the P8201 is shown in Fig. 7.2. The size of roadways is 3.9 × 5.6 m (height × width), and the roadways are located on two sides of the P8201. The roof of the roadways is supported by rockbolts and anchor cables, and steel mesh and rockbolts are installed on the ribs of the roadways. There are 14 faults in the P8201 longwall face, and there are 4 faults with fault throws greater than 10 m, as shown in Fig. 7.2. Roadways are surrounded by solid coal adjacent to normal fault F5 in the north and F4 in the south, with fault throws of 30 m and 39 m, respectively. Normal faults F2 and F6 are located on the paths of the headgate and tailgate, with fault throws of 15 m and 14 m, respectively.

### 7.2.2 Instability Characteristics of Roadway Surrounding Rock

1. Monitoring borehole layout
   When the tailgate reveals fault F6, the surrounding rock cracks develop, and water drenching will occur in the roadway. To fully understand the process of RSR failure, boreholes are set in the direction perpendicular to the tailgate roof and inclined to the tailgate rib. Three monitoring points are arranged on the tailgate roof to record the tailgate roof deformation, as shown in Fig. 7.3.
2. Analysis of monitoring results
   Figure 7.3 shows the roof deformation curve in the process of RSR failure. Among these monitoring points, monitoring point A is located in the side roof of the coal seam, and measuring points B and C are located in the fault zone. After roadway excavation, the change in roadway roof subsidence can be divided into three phases. The first phase is mainly 0 h after roadway excavation. During this stage, the displacement of the roadway roof slowly increases, and roadway roof subsidence for approximately 10% of the total subsidence. The second phase is 25–90 h after roadway excavation, and the roof subsidence increases linearly in

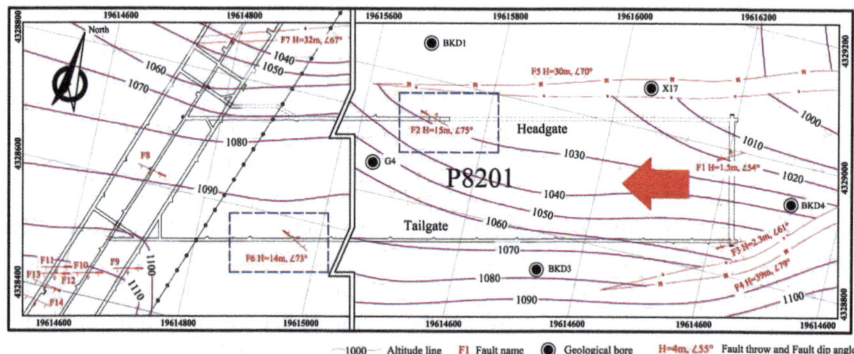


**Fig. 7.2** Longwall face and the fault layout

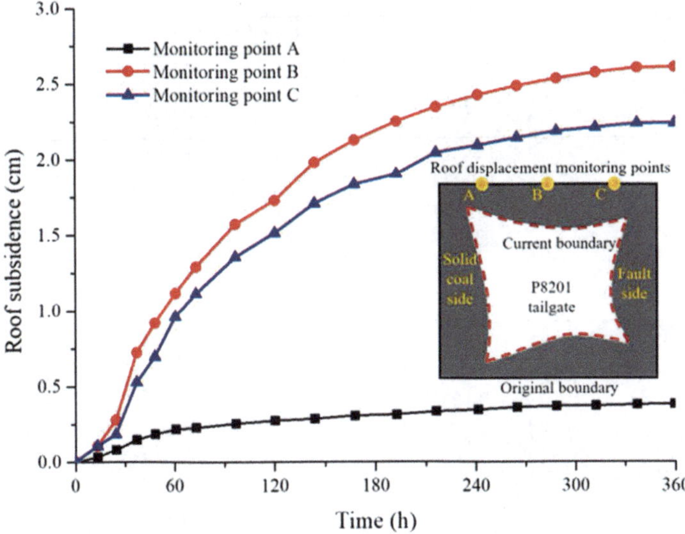


**Fig. 7.3** Roof subsidence in the process of RSR failure

this phase. The displacements at monitoring points B and C are 6.8 times and 5.8 times the displacement at monitoring point A, respectively, indicating that the fault zone is more seriously affected by excavation. The third phase is 90–360 h after roadway excavation, when the roof subsidence velocity decreases gradually, and the roof subsidence tends to be stable.

Figure 7.4 shows some observation boreholes in the tailgate of the P8201 longwall face. In the process of observation, there are many fractures and separation in inclined boreholes, and the rock mass around the borehole is seriously broken (Fig. 7.4a). Unstable loose stones in the fault and roof separation can be seen in the vertical borehole 4 m away from the tailgate roof. With increasing borehole height, weakening surfaces appear in sandstone strata, and vertical and horizontal fractures occur in the borehole wall (Fig. 7.4b). The coal on the fault side is smashed and has a large horizontal deformation, causing the metal mesh to tear and some rockbolts to fall off. At the same time, the steel strip is bent due to the continuous increase in roof subsidence, and the roadway roof begins to seep water, as shown in Fig. 7.8c.

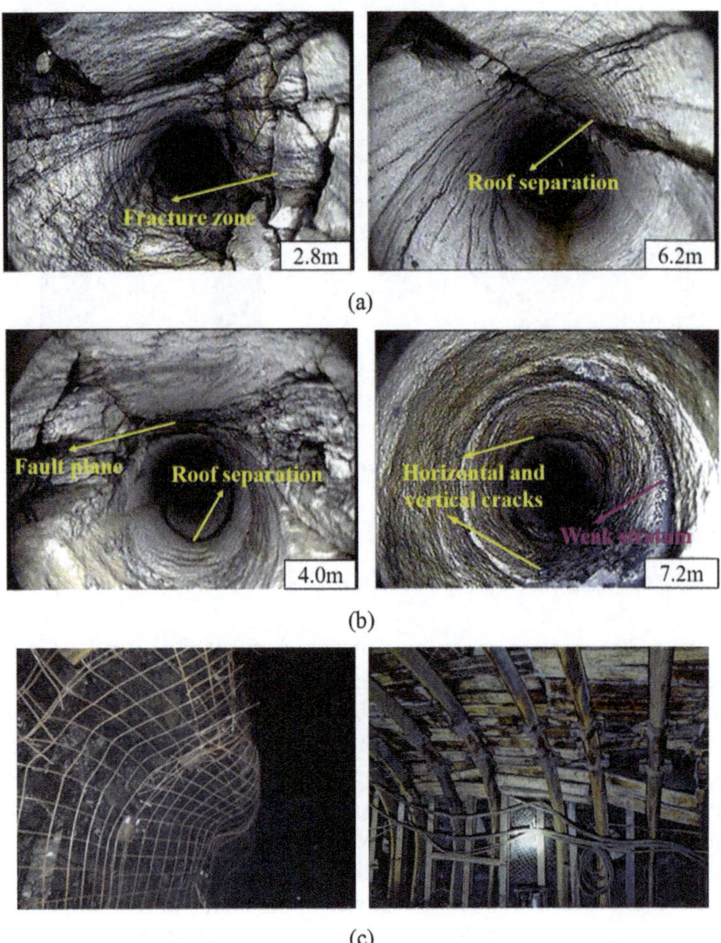


**Fig. 7.4** (**a**) Horizontal borehole observation diagram, (**b**) Vertical borehole observation diagram and (**c**) Instability of the P8201 roadway

## 7.3 Numerical Model and Parameters Determination in Engineering

### *7.3.1 Numerical Model Construction*

A UDEC-Voronoi model is established to simulate the progressive failure process of water invasion weakening and the change in mechanical behavior of RSR, as displayed in Fig. 7.5. Due to the two-dimensional simulation used in this study, the most unstable section during roadway excavation, i.e., section B in Fig. 7.5a, where the roadway penetrates the fault, is selected for simulation. The size of the model is

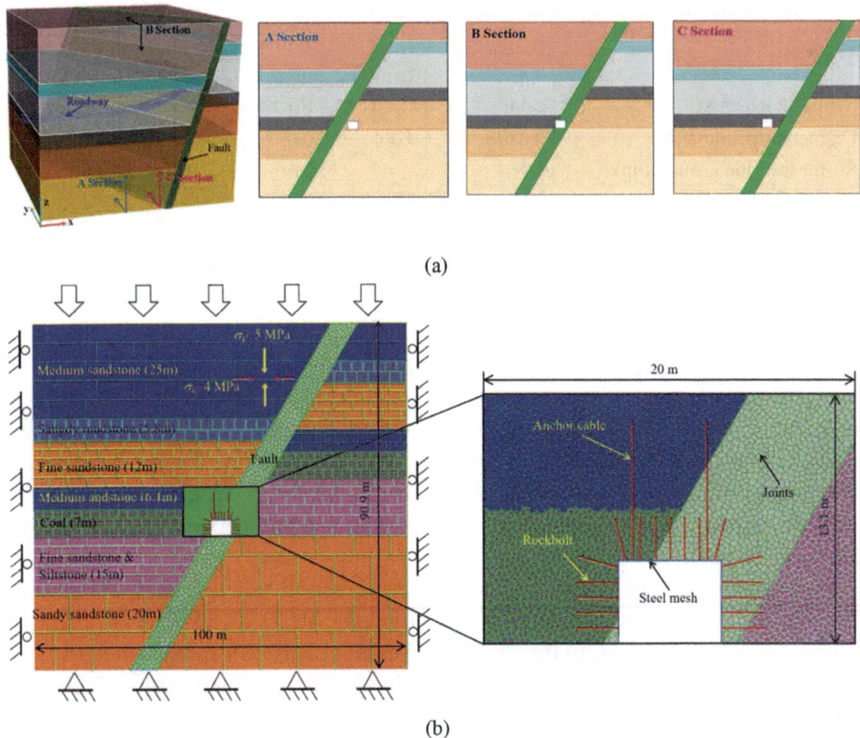


**Fig. 7.5** Model geometry and boundary conditions. (**a**) Geometric position of the model and (**b**) numerical models and boundary conditions

100 × 90.9 m (length × height). To study the failure process of the RSR, an irregular block with a side length of 0.4 m is set in the monitoring area, and the model outside the monitoring area is divided into rectangular blocks considering the model calculation efficiency.

According to the field measurement result of the in situ stress at four places (Fig. 7.2), the minimum principal stress and the maximum principal stress are perpendicular and parallel to the excavation direction of roadways, respectively. The moderate stress is equivalent to the load of strata above the coal seam, and its direction is vertical to the coal seam. The horizontal-to-vertical-stress ratios are $k_{max} = 1.5$ and $k_{min} = 0.8$. A vertical load of 5 MPa is applied to the top boundary of the model, and the horizontal displacement of on side boundaries and the vertical displacement of the bottom boundary are fixed.

In the numerical model, blocks in each rock stratum adopt the isotropic elastic strength criterion, so the block will only produce deformation without damage or instability. Coulomb slip with a residual strength criterion is used for the contacts between the blocks. Based on the stress and motion state between the blocks, contact plane shear and tension instability occur. According to the measured water

**Table 7.1** Parameters of the support elements of the P8201 roadway

| Types | Primary support | | | Strengthening support | |
|---|---|---|---|---|---|
| Support elements | Anchor cable | Steel mesh | Rockbolt | Anchor cable | Steel beam |
| Diameter/thickness (m) | 0.0178 | 0.005 | 0.022 | 0.022 | 0.11 |
| Stiffness of the grout (N/m) | 2e9 | – | 2e9 | 2e9 | – |
| Elastic modulus (GPa) | 200 | 180 | 200 | 200 | 200 |
| Tensile yield strength (kN) | 870 | – | 390 | 870 | – |
| Cohesive capacity of the grout (N/m) | 4e5 | – | 4e5 | 4e5 | – |

pressure of the roof aquifer, the pore pressure of the aquifer is set to 2 MPa, and the pore pressure of the aquifer is assumed to remain stable. After the initial stress balance, the roadway is excavated, and a pore pressure of 0 MPa is applied to the roadway boundary. At the same time, the "cable" element and "beam" element are set around the roadway to simulate the rockbolt, anchor cable, and metal mesh of the site, respectively, as shown in Fig. 7.5. The specific support parameters are shown in Table 7.1.

In the numerical simulation, roadway excavation will lead to a large unbalanced force on the excavation boundary, and the dynamic stress disturbance will lead to large-scale instability around the roadway (Sun et al., 2021a). In the actual excavation process, the surrounding rock of the roadway will form static stress. Therefore, the boundary stress release method is used to simulate roadway excavation. In this method, the stress at the boundary of the roadway is divided into 10 stages from 9.5 (in-situ stress) to 0 MPa, each of which is decreased by 10% of in situ stress, and enough numerical steps are computed to guarantee that the RSR is in a static stress environment.

### 7.3.2 Parameter Calibrations

The acquisition of rock mechanical parameters is usually measured by experimental tests, and there are differences in mechanical characteristics between intact rocks and rock masses. Therefore, Zhang and Einstein (2004) and Zhang et al. (2019c) combined an RQD index to propose a conversion formula for the mechanical parameter of intact rocks and rock masses:

$$\frac{E_{\mathrm{m}}}{E_{\mathrm{r}}} = 10^{0.0186RQD-1.91} \tag{7.1}$$

$$\frac{\sigma_{cm}}{\sigma_{cr}} = 10^{0.013RQD-1.34} \tag{7.2}$$

**Table 7.2** The mechanical characteristics of intact rocks and rock masses

| | Intact rock characteristics | | | | Rock mass characteristics | | |
|---|---|---|---|---|---|---|---|
| Lithology | $E_r$ (GPa) | $\sigma_{cr}$ (Mpa) | $T_r$ (MPa) | RQD (%) | $E_m$ (GPa) | $\sigma_{cm}$ (Mpa) | $T_m$ (MPa) |
| Sandy mudstone | 12.6 | 91.3 | 3.0 | 86 | 6.2 | 54.8 | 1.8 |
| Fine sandstone | 8.4 | 75.3 | 1.9 | 90 | 4.9 | 50.9 | 1.3 |
| 2#Coal | 3.3 | 27.5 | 1.2 | 91 | 2.0 | 19.2 | 0.8 |
| Siltstone | 18.0 | 121.2 | 3.6 | 92 | 11.4 | 87.0 | 2.6 |

**Table 7.3** Calibrated mechanical parameters of rock mass

| | Matrix properties | | | Contact properties | | | | |
|---|---|---|---|---|---|---|---|---|
| Lithology | Bulk modulus (GPa) | Shear modulus (GPa) | Density (kg/m³) | Bulk modulus (GPa) | Shear modulus (GPa) | Internal friction angle (°) | Cohesion (MPa) | Tensile strength (MPa) |
| Sandy mudstone | 14.6 | 8.5 | 2400 | 4050 | 1608 | 42(32)[a] | 8.1(0.81)[a] | 1.8 |
| Fine sandstone | 8.1 | 3.7 | 2350 | 3820 | 1456 | 40(30)[a] | 10.2(1.02)[a] | 1.3 |
| 2#coal | 3.9 | 2.8 | 1400 | 1500 | 600 | 37(27)[a] | 2.1(0.21)[a] | 0.8 |
| Siltstone | 18.9 | 13 | 2500 | 5812 | 2345 | 42(32)[a] | 12.8(1.28)[a] | 2.6 |
| Fault | 3.4 | 1.2 | 2500 | 1200 | 480 | 30(20)[a] | 1(0.1)[a] | 0.2 |

[a]Numbers in the parentheses are residual values

$$\frac{T_m}{T_r} = 10^{0.013RQD-1.34} \tag{7.3}$$

where $E_m$ and $E_r$ are the moduli of rock masses and intact rocks, respectively. $\sigma_{cm}$ and $\sigma_{cr}$ are the uniaxial compressive strengths (UCS) of rock masses and intact rocks, respectively. $T_m$ and $T_r$ are the tensile strengths of rock masses and intact rocks, respectively. The mechanical properties of the coal and rock mass after calibration are displayed in Table 7.2.

Therefore, a UCS test model with a size of 5 × 10 m (diameter × height) and a Brazilian split test model with a radius of 2.5 m are established to ensure that the model size is consistent with the roadway size. At the same time, the grid size is consistent with the roadway model to avoid the effect of grid size on the model. According to the mechanical characteristics of calibrated rock masses (UCS, tensile strength, and Young's modulus), the model's internal friction angle and cohesion were decided by a trial-and-error optimization algorithm (Gao & Stead, 2014). Among them, the mechanical parameters of the fault refer to the previous research (Hao & Azzam, 2005; Cheng et al., 2019). The mechanical parameters of the rock mass after calibration are displayed in Table 7.3.

Figure 7.6 displays the comparison between the field roof subsidence and the numerical result in the P8201 tailgate. The numerical simulation result of roof subsidence is consistent with the field result, which proves the reasonable parameter

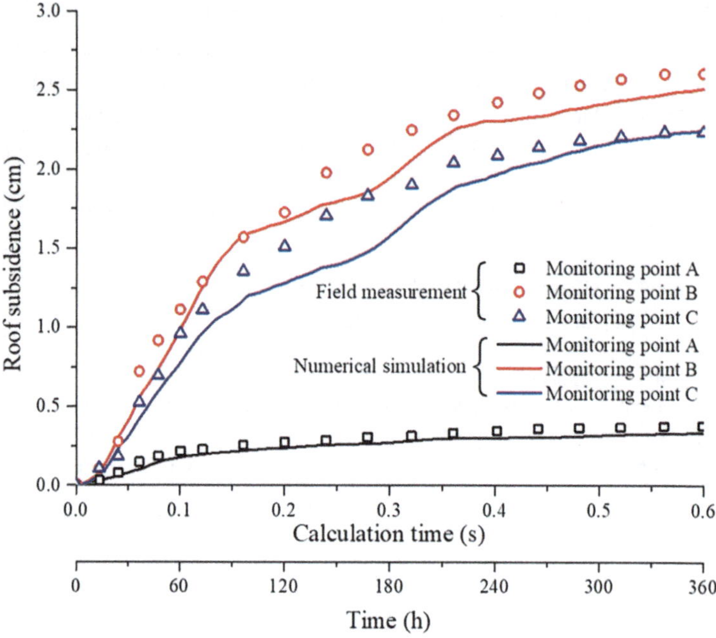


**Fig. 7.6** Comparison of numerical roof subsidence and field results in P8201 tailgate

selection and the applicability of the water immersion weakened simulation method to research the progressive failure characteristics of the RSR. The numerical simulation is unable to effectively simulate the maximum separation between blocks within rock strata, resulting in smaller simulation results than those measured in the field. It is noteworthy that the UDEC calculation time is not equal to the field measurement time, and there is a linear correlation between them. The subsequent time in this chapter is based on the calculation time.

## 7.4 Failure Analysis of Water Invasion Weakening of Roadway Surrounding Rock

When roadway excavation is close to the fault, the stress redistribution caused by the roadway excavation and roof water pressure will lead to fault activation, and the fault state between the upper and lower plates may change from "adhesion" to "tension or shear," providing a seepage channel for roof water to flow into the roadways. Therefore, the influence of rock mass strength weakening around the seepage channel should be considered in the process of analyzing the instability of RSR. Figure 7.7 shows the variation curve of the RSR damage and roof subsidence (monitoring point B) with roadway excavation time in the monitoring area. According to the damage evolution process of the RSR, the damage development mainly includes four stages (Stage A, Stage B, Stage C and Stage D are defined as shown in Fig. 7.7).

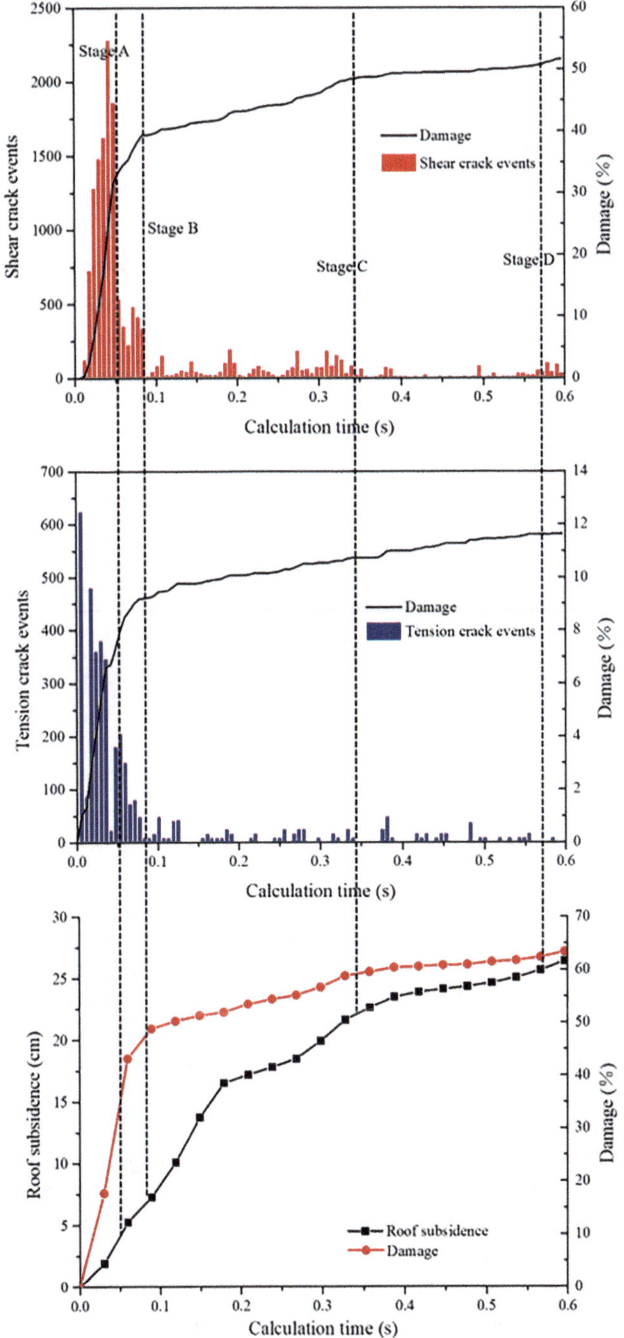


**Fig. 7.7** Microcrack event, damage ratio, and roof subsidence during roadway instability

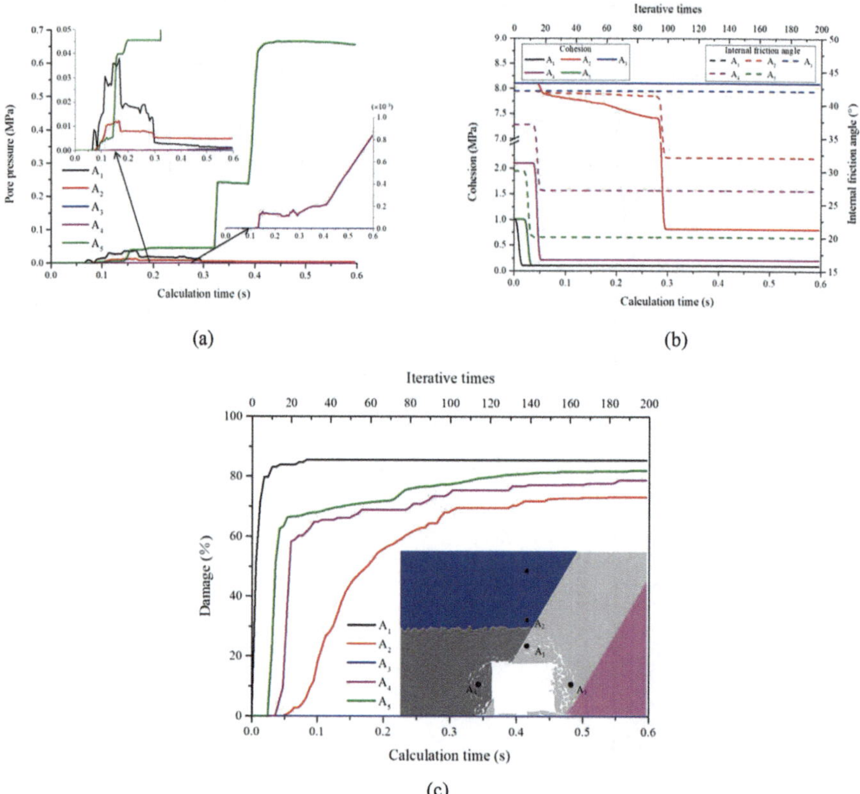


**Fig. 7.8** Pore pressure, strength weakening, and damage ratio in the RSR with various calculation time (various iterations)

Based on the fluid-solid simulation method of the RSR stability study under water immersion weakening, the water immersion weakening parameter, pore pressure, and damage ratio with various calculation time are monitored and analyzed, as displayed in Fig. 7.8. The monitoring arrangement of the RSR is shown in Fig. 7.8c.

According to Sect. 7.3, the moisture content and permeability of the surrounding rock are related only to the contact aperture. Figure 7.8 shows that there are some differences in the weakening parameters of each monitoring point in RSR, which induces the damage ratio to be different. At the initial stage of roadway excavation (stage A), the pore water pressure of all monitoring points is 0 MPa, indicating that there is no weakening effect at this stage. The cohesion and internal friction angle at monitoring points $A_1$, $A_4$, and $A_5$ directly decreased to residual strength, resulting in a sharp increase in the damage ratio at $A_1$, $A_4$, and $A_5$, which increased by 58.3%, 62.5%, and 79.7%, respectively. When the calculation time increased to 0.07 s (at stage B), the pore pressure at monitoring points $A_1$ and $A_2$ began to increase, and $A_1$ was closer to the seepage channel formed by the activation of the fault, resulting in a higher pore pressure at $A_1$ than at $A_2$. Under the effect of water, the mechanical parameters at monitoring point $A_2$ began to weaken, and the damage ratio at $A_2$ slowly increased to 2.9%. With increasing calculation time, the internal friction angle and cohesion of monitoring point $A_2$ finally decreased to residual strength, while the water pressure of monitoring point $A_3$ remained at 0 MPa. The results show that the water pressure has not infiltrated into $A_3$ during the simulation period. In addition, as the calculation time increased, the pore pressure at monitoring point $A_5$ increased to 0.68 MPa and then remained stable, indicating that fractures at measuring point $A_5$ are not connected to a roadway.

To further analyze the surrounding rock instability process, the distribution of cracks, maximum principal stress, and pore pressure in four stages is shown in Figs. 7.9, 7.10, and 7.11. At the beginning of stage A, the contacts are mainly tensile cracks. However, shear failure becomes the main damage form (Fig. 7.7). As shown in Fig. 7.10a, the peak values of concentrated stress in the two upper corners of the tailgate are 1.4 and 2.1 MPa, respectively, and the stress concentration on the fault rib is higher than that on the coal seam rib. Due to the concentrated stress, many microscopic cracks are generated around the tailgate, and roof subsidence slowly increases in stage A. During this period, monitoring points $A_2$ and $A_3$ are in the elastic zone, while the contacts at $A_1$, $A_4$, and $A_5$ are unstable, which are in the form of microscopic cracks, as shown in Fig. 7.9a. Because the flow of liquid in the elastic zone is very slow, no water seepage occurs in the tailgate (Fig. 7.11a).

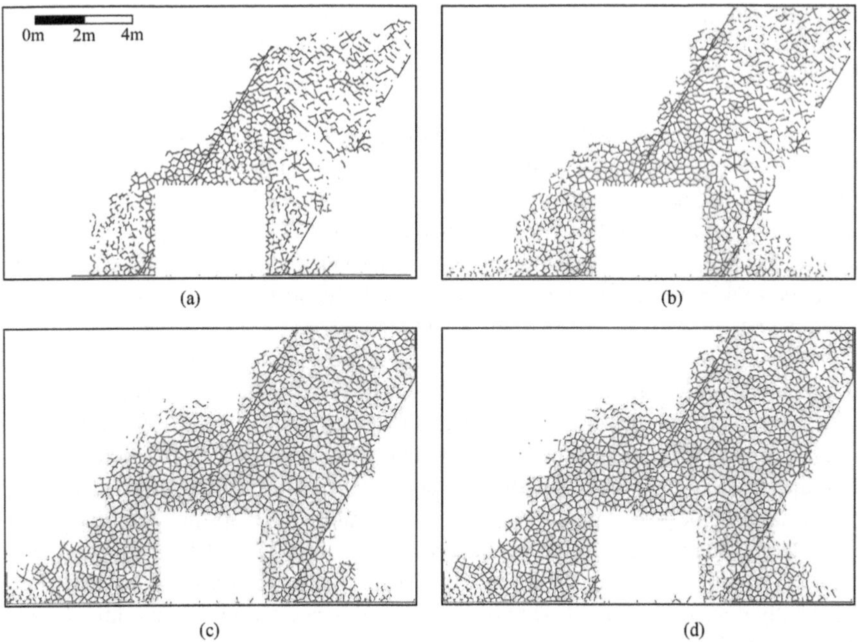


**Fig. 7.9** The development process of contact cracks (tensile cracks in red and shear cracks in blue) in tailgate surrounding rock under the condition of water invasion weakening. (**a**) Stage A, (**b**) Stage B, (**c**) Stage C, and (**d**) Stage D

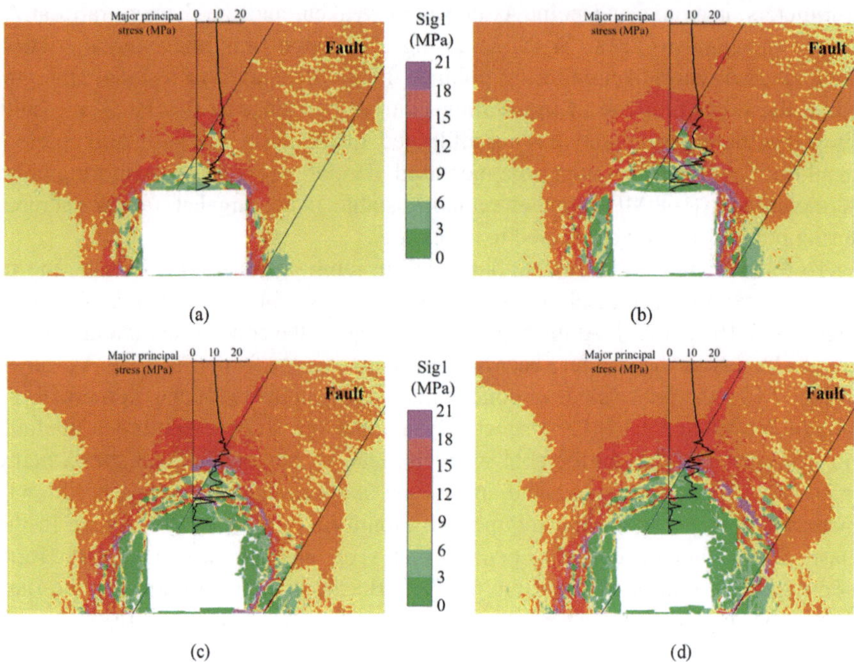


**Fig. 7.10** The distribution variation of maximum principal stress in tailgate surrounding rock under the condition of water invasion weakening. (**a**) Stage A, (**b**) Stage B, (**c**) Stage C, and (**d**) Stage D

At stage B, the contact damage ratio in the monitoring area increased from 36.3% to 51.5% (Fig. 7.10), and roof cracks developed into sand-mudstone strata. At this time, the fault is activated, the RSR cracks connect the aquifers, the pore pressure of monitoring points $A_1$ and $A_2$ begins to increase, and the seepage channel of the roof aquifer has been formed, as shown in Fig. 7.11b. Due to the weakening effect of the coal and rock mass around the seepage channel, the destressed zone of the tailgate roof extends upward by 1.2 m (Fig. 7.10b). The degree of stress concentration on the tailgate ribs has not changed, indicating that the tailgate surrounding rock is still dominated by microscopic cracks, as shown in Fig. 7.10b. In this stage, the roof subsidence velocity increases from 1.1 cm/h in stage A to 1.8 cm/h, and water seepage occurs in the tailgate roof. The growing roof subsidence velocity demonstrates that the roadway is in the unsteady state and local collapse is likely to occur.

At stage C, the position of roof stress concentration develops into the sandy mudstone strata, and the peak value of concentrated stress is 18.2 MPa (2.6 times higher than the in situ stress), as shown in Fig. 7.10c. Since the cracks in the RSR have been fully developed in the first two stages, the damage ratio in the monitoring area has increased only by 9.1%. The broken stratum aggravates the subsidence of the rock beam near the roof, which leads to the separation of the rock strata. The roof subsidence increased from 7.28 to 22.6 cm, an increase of 210.6% compared with

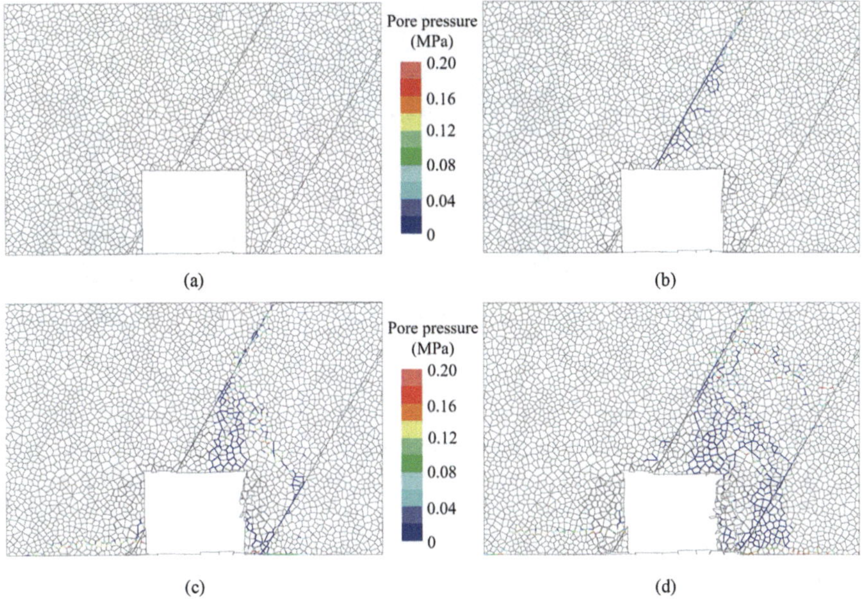


**Fig. 7.11** Pore pressure change process in tailgate surrounding rock under the condition of water invasion weakening. (**a**) Stage A, (**b**) Stage B, (**c**) Stage C, and (**d**) Stage D

the previous two stages, but the roof subsidence velocity gradually decreased (Fig. 7.7). As shown in Fig. 7.11c, the seepage path in the fault is divided into two parts, one of which infiltrates into the tailgate along the roof. Among them, when the microscopic cracks at monitoring points $A_1$ and $A_2$ expand into macroscopic cracks, the pore pressures at $A_1$ and $A_2$ gradually decrease (Fig. 7.8a). The other part infiltrates into the interface between the fault zone and lower plate. The failure form at monitoring point $A_5$ is mainly microcracks and no seepage channel is formed, resulting in an S-shaped increase in pore pressure at $A_5$.

At stage D, the weakening process of the RSR continues, and the destressed zone of the surrounding rock further extends to both ribs of the tailgate, as shown in Fig. 7.9d. During this period, the degree of crack development and the roof subsidence in the monitoring area increase slowly. Figure 7.8 shows that the shear damage ratio plays a dominant role in the failure process of the RSR, and the shear damage ratio is 4.5 times the shear damage ratio of the tensile damage ratio (Fig. 7.7).

As displayed in Fig. 7.12, the damage ratio of RSR in the anchorage zone of the tailgate is more than 65%, and the damage at the rockbolt end is the largest. In the fault zone, except for the large damage ratio near the fault, the damage ratio in other parts is relatively small (40 and a large damage ratio occurs mainly in the area near the seepage channel (Fig. 7.11d). At stage D, the maximum subsidence of the tailgate roof is 26.4 cm, which indicates that the existing support can no longer meet the stability of RSR.

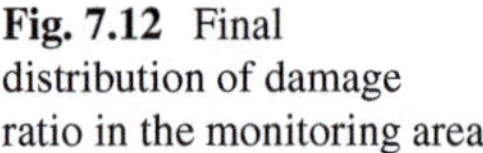

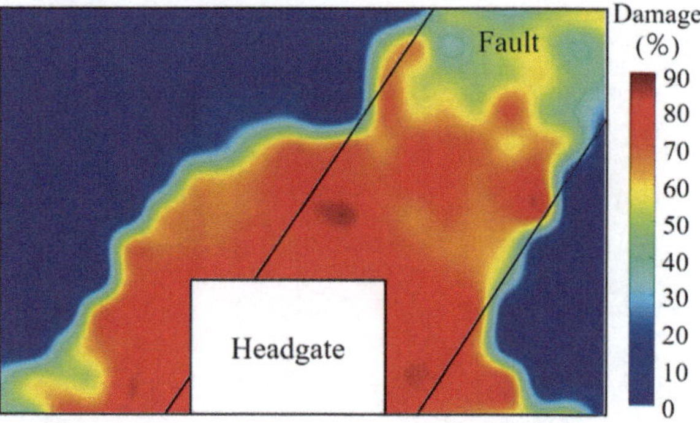


**Fig. 7.12** Final distribution of damage ratio in the monitoring area

The simulation result indicates that the mechanical strength of coal and rock masses decreases after water invasion, which causes stress redistribution to transfer to the roadway periphery and promotes crack development in RSR. Under the effect of long-term water immersion, the roadway will eventually become unstable. Therefore, roof dewatering can be used to maintain RSR stability. And meanwhile, the reinforced support is applied to control RSR deformation.

## 7.5 Control Techniques

A numerical simulation study shows that water drenching of the roof not only reduces the support strength but also weakens the mechanical strength of rock masses (Showkati et al., 2021). Roof dewatering treatment has been proven to be an effective control method to reduce the damage of RSR. Dewatering holes are drilled near the fault of the headgate, as shown in Fig. 7.13. Based on the simulation results in Sect. 7.4, it is clear that seepage channels only form near the fault zone. Therefore, the pore pressure of the aquifer is set to 0 MPa in the numerical simulation to enable the drainage process during the simulation.

In the case of hydrophobic drilling in the roof, weakening of the RSR strength can be avoided, and the RSR stability can be enhanced. Figure 7.14 shows the mechanical response of the RSR in stage D after roof dewatering. As shown in Fig. 7.14a, the roof subsidence shows an obvious asymmetry. Compared with the roof water-rich condition, the maximum subsidence is reduced from 26.4 to 18.77 cm, a decrease of 28.9%. The destressed zones on the left and right ribs of the roadway are 1 and 1.2 m, respectively, which are 0.5 and 0.8 m less than the destressed zones under weakening conditions. The height of the roof destressed zone is the same, but the range of the roof destressed zone is reduced. The damage distribution of the RSR in Fig. 7.14b is different from the damage distribution in Fig. 7.12. Under the condition of roof dewatering, the damage ratio of the coal seam rib is less than 65%; the area with greater damage (70–80%) is concentrated mainly in the fault zone, and the damage range of RSR is reduced. The results show that the

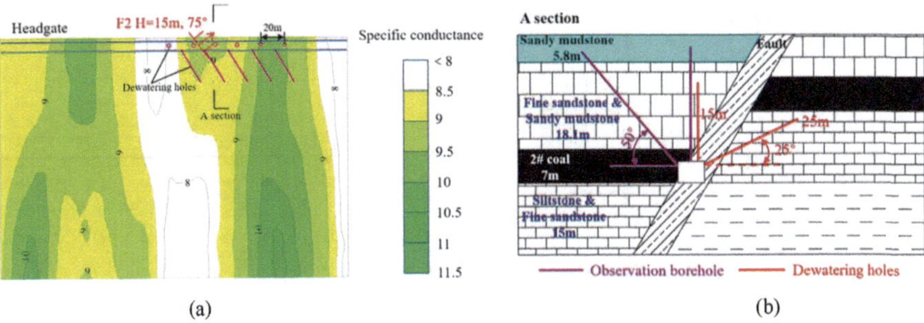


**Fig. 7.13** Layout of boreholes in the headgate. (**a**) Planar graph and (**b**) cross-section graph in the A section location

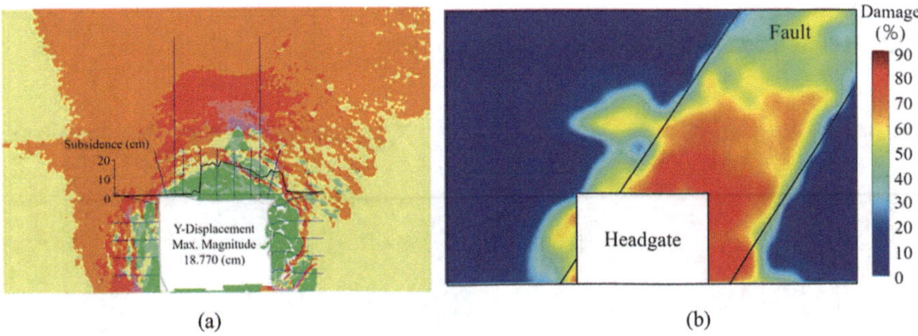


**Fig. 7.14** The mechanical response of RSR in stage D after roof dewatering treatment. (**a**) Maximum principal stress distribution and (**b**) damage distribution

roof dewatering scheme does not prevent the activation of fault around the roadway. Although the roof dewatering scheme can remove the threat of roof water drenching, the rockbolts in the fault zone have lost their bearing characteristics. There is a high possibility of roof collapse and roadway rib spalling, which requires higher support strength.

To decrease the RSR deformation, strengthening support is carried out for the structures of roadways prone to instability. Through the above analysis, it can be seen that the roof subsidence of the fault rib is higher than that on the coal seam rib (Fig. 7.3). The weak structure of the roadway is the fault-side roof and rib. According to the asymmetry control measures of "the overall RSR stability, strengthening support of weakness structures" (Xu et al., 2021), three anchor cables and steel beam structures are added to the fault-side roof. The strengthening support parameters are shown in Table 7.1. Among these cables, the outermost anchor is inclined outward by 15°, and the middle anchor cable is perpendicular to the roof. In addition, the bottom of the anchor cables is fixed on the steel beam. Two anchor cables are added to the fault-side rib, both of which are perpendicular to the roadway rib, and the ends of anchor cables are located in the elastic zone, as displayed in Fig. 7.15.

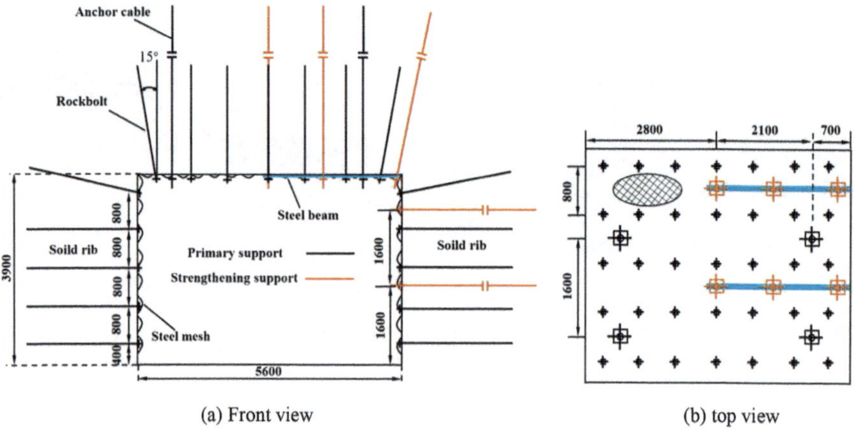


**Fig. 7.15** Support scheme. (**a**) Front view, (**b**) top view

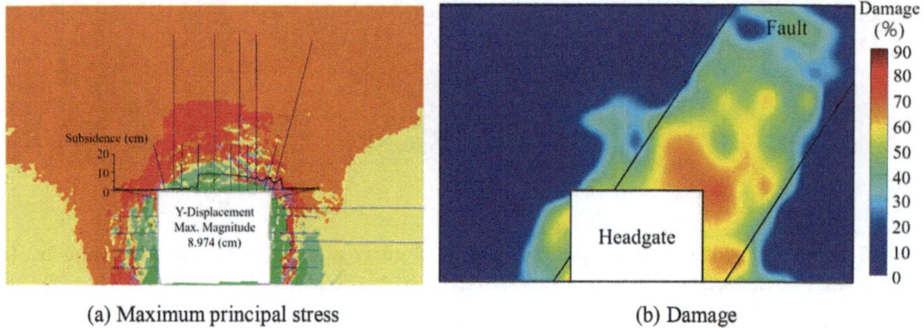


**Fig. 7.16** The mechanical response of RSR in stage D after roof dewatering treatment and roadway strengthening support. (**a**) Maximum principal stress, (**b**) Damage

During the simulation, the reinforced support is supported immediately after the excavation of a roadway.

By roof dewatering treatment and strengthening the support, the development of cracks in the RSR can be efficiently controlled, so as to realize the purpose of compatible deformation of the RSR, as shown in Fig. 7.16. Compared with only considering the roof dewatering condition, the maximum subsidence is reduced from 18.77 to 8.97 cm, a decrease of 52.2%. The range of the roof destressed zone is reduced from 3.1 to 1.4 m, and rockbolts on the fault-side rib still have bearing capacity. The results show that the development of macroscopic cracks is also significantly restrained by anchor cables, resulting in some rock bridges between cracks. These rock bridges greatly reduce the interaction between cracks and improve the stability of rock masses in the anchorage zone. Figure 7.16b clearly shows that strengthening support does not prevent the activation of faults or the formation of seepage channels. After strengthening support, the damage

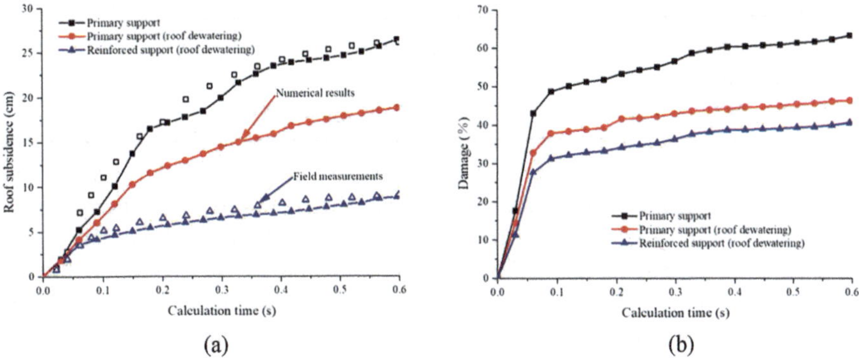


**Fig. 7.17** Under different measures during the failure process of RSR, the change curve of (**a**) subsidence at monitoring point B and (**b**) damage ratio

distribution range in the monitoring area is consistent with the damage distribution range of the roof dewatering scheme. However, the maximum damage in the fault zone is 60–65%, which is 15–20% less than the maximum damage of the roof dewatering scheme.

Figure 7.17 indicates the change curve of subsidence at monitoring point B and the damage ratio in the monitoring area with increasing calculation time under different measures. On the whole, the numerical results are consistent with the field-measured results. When the roadway roof adopts only the measure of roof dewatering, the damage ratio of the surrounding rock is controlled. Compared with the roof water-rich condition, the damage ratio decreases from 63.3% to 46.2%. However, the subsidence at monitoring point B is reduced from 26.4 to 18.77 cm, and the roof deformation in the fault zone is still very large. After roof dewatering treatment and roadway strengthening support, roadway deformation and damage ratio are controlled. Compared with the roof water-rich condition, the final subsidence at monitoring point B decreases from 26.4 to 9.0 cm, which is a decrease of 65.9%, and the damage ratio in the monitoring area decreases from 63.3% to 40.5%.

## 7.6 Conclusions

1. Based on the characteristics of water immersion weakening of rock masses, a fluid-solid coupling simulation method considering water immersion weakening is proposed. This method calculates the water content and permeability of rock masses based on the contact aperture. Through laboratory tests of coal and rock samples with different water contents, the relationship between moisture content and the weakening coefficient of mechanical parameters is obtained. On this basis, the calibrated mechanical parameters of the coal and rock mass are embed-

ded into the UDEC numerical simulation software using the FISH language, which can realize the fluid-solid coupling analysis of the RSR stability under water invasion weakening conditions.

2. The numerical simulation results reproduce the progressive failure process of roof water drenching and surrounding rock failure after roadway excavation. The essence of progressive failure is that the seepage channel is formed due to the activation of fault, and then the crack gradually expands due to the strength deterioration of the RSR. Under the primary support conditions, the maximum subsidence of the tailgate roof is 26.4 cm, and the damage ratio in the study area is 63.3%. Among them, the shear damage ratio plays a dominant role in the failure process of RSR, and the shear damage ratio is 4.5 times the shear damage ratio of the tensile damage ratio.
3. Considering the actual site condition and simulation result, this chapter puts forward dewatering of the roof aquifer near fault F2 of the P8201 headgate, so as to avoid the occurrence of roof water drenching events. And meanwhile, the reinforced support is applied to control the damage and deformation of the RSR. After adopting the above measures, compared with the roof water-rich condition, the roof subsidence and surrounding rock damage of the headgate are reduced by 65.9% and 22.8%, respectively.

# Chapter 8
# Conclusion

This book has systematically addressed the critical challenge of water-rock interaction (WRI) in underground coal mining, presenting a logical and progressive pathway from fundamental problem identification to advanced predictive simulation. The preceding chapters are not isolated studies but interconnected components of an integrated framework for analyzing, understanding, and mitigating WRI-related hazards. This concluding chapter synthesizes these components, highlighting how each chapter builds upon the last to form a comprehensive methodology for tackling WRI problems in both academic research and engineering practice.

## 8.1 The Integrated WRI Analysis Framework

The logical flow of this book can be visualized as a cohesive analytical framework, progressing from the macroscopic to the microscopic, and back to the macroscopic engineering scale through numerical modeling:

Problem Definition (Chap. 1): The foundation is laid by identifying and categorizing the primary WRI challenges in coal mining: water inrush from geological structures, stability issues in roadways and coal pillars due to water weakening, and the emerging focus on utilizing abandoned mines and underground reservoirs. This chapter establishes the why—the critical engineering problems that demand solutions. It also reviews the two primary analysis tools: laboratory testing and numerical simulation, setting the stage for the detailed investigations that follow.

Macroscopic Weakening Phenomenon (Chap. 2): Building on the problem definition, Chap. 2 delves into the what—the macroscopic manifestation of WRI. It comprehensively reviews how water content, immersion time, and wetting-drying cycles degrade the mechanical properties (strength, modulus, brittleness) of coal and rock. This chapter establishes the empirical evidence for water weakening, which is the primary concern for engineering stability.



C. Zhang, *Water Rock Interaction in Underground Coal Mining*,
https://doi.org/10.1007/978-981-95-9957-8_8

Micromechanism and Characterization (Chap. 3): To understand how behind the macroscopic weakening, Chap. 3 advances the analysis to the microscopic scale. It introduces a novel, accurate method for characterizing the internal fabric of coal—pores and various minerals—using CT scanning calibrated with NMR and XRD. This chapter provides the crucial link between the observable mechanical degradation and the internal structural evolution, revealing that weakening is primarily driven by the dissolution and expansion of hydrophilic minerals (e.g., clay), which alters pore connectivity and fracture networks.

Long-Term Evolution (Chap. 4): Chapter 4 extends the temporal scope, moving from short-term saturation effects to the long-term evolution of rock properties. Through a meticulous experimental study on sandstone, it quantifies how prolonged immersion leads to continuous physicochemical reactions (mineral dissolution), further altering the pore structure and leading to a transition in failure mode from brittle to more ductile. This chapter provides essential data on the time-dependent nature of WRI, which is critical for assessing the long-term stability of structures like coal pillar dams in underground reservoirs.

Simulation Bridge to Engineering Scale (Chap. 5): The insights from laboratory-scale experiments (Chaps. 2, 3, and 4) are of limited direct use without a method to apply them to large-scale engineering problems. Chapter 5 builds the vital bridge by proposing fluid-solid coupling simulation methods (for both Finite Element and Discrete Element models) that incorporate real-time updates for permeability (based on stress/damage) and strength (based on water content). This chapter translates the qualitative and quantitative understanding from the lab into a quantitative, predictive numerical tool.

Engineering Validation and Application (Chaps. 6 and 7): The framework is validated and applied through two detailed case studies. Chapter 6 demonstrates the application of the simulation method to predict the progressive failure of a residual coal pillar under the coupled impact of mining stress and water invasion from an underground reservoir. Chapter 7 applies the framework to analyze the instability of a roadway crossing a water-rich fault zone. These chapters confirm the practical value of the integrated approach, showing how it can be used to diagnose failure mechanisms, evaluate control measures (e.g., optimal pillar size, dewatering strategies, and reinforced support), and inform engineering decisions.

## 8.2 Synergistic Contributions of the Framework

The strength of this framework lies in the synergy between its components: Chaps. 1, 2, 3, and 4 provide the essential physical parameters and constitutive relationships (e.g., strength vs. water content, permeability vs. stress) required to calibrate the numerical models developed in Chap. 5. The advanced characterization techniques in Chap. 3 offer a visual and quantitative explanation for the mechanical behaviors observed in Chaps. 2 and 4, moving beyond empirical correlation to mechanistic understanding. The predictive simulations in Chaps. 5, 6, and 7 allow for the "scale-up" of laboratory results, enabling the analysis of complex, boundary-specific engineering scenarios that are impossible to replicate physically.

## 8.3 Future Outlook

While this book presents a significant step forward, the framework also highlights avenues for future research: Long-Term Data: There is a pressing need for more laboratory and field data on the weakening effects of WRI over timescales of years or decades, particularly for broken rock masses in goaf areas. Advanced Simulation: The framework can be enhanced by incorporating more sophisticated simulation techniques, such as fully coupled hydro-mechanical-chemical models, and by extending the methods to 3D Discrete Element and Finite-Discrete Element methods to better capture complex fracture networks. Standardization and Automation: The micro-characterization methods could be further developed into more standardized and automated procedures for wider application. Broader Applications: The integrated framework is not limited to coal mining. It holds significant potential for application in other fields such as slope stability, tunnel engineering in water-rich strata, and the stability of reservoir dams, where water-rock interactions play a decisive role.

In conclusion, this book has constructed a robust, integrated framework that seamlessly connects the identification of WRI problems, the fundamental investigation of weakening mechanisms, and the practical application of predictive models for engineering-scale analysis. By providing both a deep scientific understanding and practical engineering tools, it aims to equip scientists and engineers with the knowledge and methodology needed to enhance safety, efficiency, and sustainability in underground coal mining and beyond.

# References


Aggelis, D. G., Mpalaskas, A. C., Ntalakas, D., & Matikas, T. E. (2012). Effect of wave distortion on acoustic emission characterization of cementitious materials. Construction and Building Materials, 35, 183–190.

Ai, T., Wu, S., Zhang, R., Gao, M., Zhou, J., Xie, J., & Zhang, Z. (2021). Changes in the structure and mechanical properties of a typical coal induced by water immersion. *International Journal of Rock Mechanics and Mining Sciences, 138*, 104597.

Al-Bazali, T. M., Al-Mudh'hi, S., & Chenevert, M. E. (2011). An experimental investigation of the impact of diffusion osmosis and chemical osmosis on the stability of shales. Petroleum Science and Technology. 29(3), 312–323.

Alber, M., Fritschen, R., Bischoff, M., & Meier, T. (2009). Rock mechanical investigations of seismic events in a deep longwall coal mine. *International Journal of Rock Mechanics and Mining Sciences, 46*(2), 408–420.

Al-Shaaibi, S. K., Al-Ajmi, A. M., & Al-Wahaibi, Y. (2013). Three dimensional modeling for predicting sand production. *Journal of Petroleum Science and Engineering, 109*, 348–363.

An, W., Wang, L., & Chen, H. (2020). Mechanical properties of weathered feldspar sandstone after experiencing dry-wet cycles. *Advances in Materials Science and Engineering, 2020*, 1–15.

Andrews, B. J., Cumberpatch, Z. A., Shipton, Z. K., & Lord, R. (2020). Collapse processes in abandoned pillar and stall coal mines: Implications for shallow mine geothermal energy. *Geothermics, 88*, 101904.

Azhar, M. U., Zhou, H., Yang, F., Younis, A., Lu, X., Fang, H., & Geng, Y. (2020). Water-induced softening behavior of clay-rich sandstone in Lanzhou Water Supply Project, China. *Journal of Rock Mechanics and Geotechnical Engineering, 12*(3), 557–570.

Babadagli, T., Ren, X., & Develi, K. (2015). Effects of fractal surface roughness and lithology on single and multiphase flow in a single fracture: An experimental investigation. *International Journal of Multiphase Flow, 68*, 40–58.

Bai, Q., & Tu, S. (2019). A general review on longwall mining-induced fractures in near-face regions. *Geofluids, 2019*, 3089292.

Bai, Q., & Tu, S. (2020). Numerical observations of the failure of a laminated and jointed roof and the effective of different support schemes: A case study. *Environment and Earth Science, 79*(10), 1–18.

Bai, H., Ma, D., & Chen, Z. (2013). Mechanical behavior of groundwater seepage in karst collapse pillars. *Engineering Geology, 164*, 101–106.

Bai, Q., Tu, S., Zhang, C., & Zhu, D. (2016). Discrete element modeling of progressive failure in a wide coal roadway from water-rich roofs. *International Journal of Coal Geology, 167*, 215–229.




C. Zhang, *Water Rock Interaction in Underground Coal Mining*,
https://doi.org/10.1007/978-981-95-9957-8

Bai, Q., Tu, S., Wang, F., & Zhang, C. (2017). Field and numerical investigations of gateroad system failure induced by hard roofs in a longwall top coal caving face. *International Journal of Coal Geology, 173*, 176–199.

Bai, Q., Tibbo, M., Nasseri, M. H. B., & Young, R. P. (2019b). True triaxial experimental investigation of rock response around the mine-by tunnel under an in situ 3D stress path. *Rock Mechanics and Rock Engineering, 52*, 3971–3986.

Bai, B., Nie, Q., Zhang, Y., Wang, X., & Hu, W. (2021). Cotransport of heavy metals and SiO2 particles at different temperatures by seepage. *Journal of Hydrology, 597*, 125771.

Barton, N., & Choubey, V. (1977). The shear strength of rock joints in theory and practice. *Rock Mechanics, 10*(1), 1–54.

Bathe, K. J. (2007). Finite element method. In B. Wah (Ed.), *Wiley encyclopedia of computer science and engineering* (pp. 1–12). Wiley.

Baud, P., Zhu, W., & Wong, T. F. (2000). Failure mode and weakening effect of water on sandstone. *Journal of Geophysical Research: Solid Earth, 105*(B7), 16371–16389.

Belytschko, T., Gracie, R., & Ventura, G. (2009). A review of extended/generalized finite element methods for material modeling. *Modelling and Simulation in Materials Science and Engineering, 17*(4), 043001.

Berger, G., Lacharpagne, J. C., Velde, B., Beaufort, D., & Lanson, B. (1997). Kinetic constraints on illitization reactions and the effects of organic diagenesis in sandstone/shale sequences. Applied Geochemistry, 12(1), 23–35.

Bhushan, B. (2000). Surface roughness analysis and measurement techniques. In *Modern tribology handbook, two volume set* (pp. 79–150). CRC Press.

Bian, Z., Zhou, Y., Zeng, C., Huang, J., Pu, H., & Axel, P. (2021). Discussion of the basic problems for the construction of underground pumped storage reservoir in abandoned coal mines. *Journal of China Coal Society, 46*(10), 3308–3318. (in Chinese).

Bilir, M. E., & Sarıgül, G. (2021). Stability assessment of Terzili tunnel in swelling rock mass. Arabian Journal of Geosciences, 14(11), 1–15.

Boral, P., Varma, A. K., & Maity, S. (2021). Nitration of jharia basin coals, India: A study of structural modifications by xrd and ftir techniques. *International Journal of Coal Science & Technology, 8*, 1034–1053.

Brantut, N., Heap, M. J., Baud, P., & Meredith, P. G. (2014). Mechanisms of time-dependent deformation in porous limestone. Journal of Geophysical Research: Solid Earth, 119(7), 5444–5463.

Brumby, P. E., Sato, T., Nagao, J., Tenma, N., & Narita, H. (2015). Coupled LBM–DEM microscale simulations of cohesive particle erosion due to shear flows. *Transport in Porous Media, 109*(1), 43–60.

Bu, W., & Xu, H. (2020). Research on the effect of dip angle on shear stress on normal fault plane and water inrush in floor strata during mining activities. *Geotechnical and Geological Engineering, 38*(5), 4407–4421.

Cai, B., Duan, G., He, C., Gao, Y., Li, Y., Xu, Z., & Jiang, W. (2018, December). New technologies enhance efficiency of hydraulic fracturing stimulation in heavy oil reservoirs. In *SPE international heavy oil conference and exhibition*. OnePetro.

Cai, X., Zhou, Z., Tan, L., Zang, H., & Song, Z. (2020a). Fracture behavior and damage mechanisms of sandstone subjected to wetting-drying cycles. *Engineering Fracture Mechanics, 234*, 107109.

Cai, X., Zhou, Z., Zang, H., & Song, Z. (2020b). Water saturation effects on dynamic behavior and microstructure damage of sandstone: Phenomena and mechanisms. *Engineering Geology, 276*, 105760.

Cao, Z., Gu, Q., Huang, Z., & Fu, J. (2022). Risk assessment of fault water inrush during deep mining. *International Journal of Mining Science and Technology, 32*(2), 423–434.

Ceccato, A., Viola, G., Antonellini, M., Tartaglia, G., & Ryan, E. J. (2021). Constraints upon fault zone properties by combined structural analysis of virtual outcrop models and discrete fracture network modelling. *Journal of Structural Geology, 152*, 104444–104462.

Chaskalovic, J. (2008). *Finite element methods for engineering sciences: Theoretical approach and problem solving techniques*. Springer.

Chen, D., Pan, Z., & Ye, Z. (2015). Dependence of gas shale fracture permeability on effective stress and reservoir pressure: Model match and insights. *Fuel, 139*, 383–392.

Chen, S., Wang, H., Wang, H., Guo, W., & Li, X. (2016). Strip coal pillar design based on estimated surface subsidence in Eastern China. *Rock Mechanics and Rock Engineering, 49*(9), 3829–3838.

Chen, G., Li, T., Wang, W., Zhu, Z., Chen, Z., & Tang, O. (2019a). Weakening effects of the presence of water on the brittleness of hard sandstone. *Bulletin of Engineering Geology and the Environment, 78*(3), 1471–1483.

Chen, S., Jiang, T., Wang, H., Feng, F., Yin, D., & Li, X. (2019b). Influence of cyclic wetting-drying on the mechanical strength characteristics of coal samples: A laboratory-scale study. *Energy Science & Engineering, 7*(6), 3020–3037.

Chen, X., He, P., & Qin, Z. (2019c). Strength weakening and energy mechanism of rocks subjected to wet–dry cycles. *Geotechnical and Geological Engineering, 37*(5), 3915–3923.

Chen, Y., Selvadurai, A. P. S., & Liang, W. (2019d). Computational modelling of groundwater inflow during a longwall coal mining advance: A case study from the Shanxi Province, China. *Rock Mechanics and Rock Engineering, 52*(3), 917–934.

Chen, F., Wang, Y., Jiang, W., & Zheng, S. (2021a). Numerical simulation of ground movement induced by water and sand gushing in subway through fault based on DEM-CFD. *Computers and Geotechnics, 139*, 104282.

Chen, P., Tang, S., Liang, X., Zhang, Y., & Tang, C. (2021b). The influence of immersed water level on the short-and long-term mechanical behavior of sandstone. *International Journal of Rock Mechanics and Mining Sciences, 138*, 104631.

Chen, J., Cheng, W. M., Guo, L. W., Zhu, L. Q., Qi, Y. P., Liu, Z., & Huang, Q. M. (2022a). Law of monitoring parameters of water injection coal based on the combined analysis of CT and ultrasonic technology. *Journal of Mining & Safety Engineering, 39*(4), 786–796.

Chen, T., Yin, H., Zhai, Y., Xu, L., Zhao, C., & Zhang, L. (2022b). Numerical simulation of mine water inflow with an embedded discrete fracture model: Application to the 16112 working face in the Binhu coal mine, China. *Mine Water and the Environment, 41*, 156–167.

Cheng, G., Li, L., Zhu, W., Yang, T., Tang, C., Zheng, Y., & Wang, Y. (2019). Microseismic investigation of mining-induced brittle fault activation in a Chinese coal mine. *International Journal of Rock Mechanics and Mining Sciences, 123*, 104096.

Cheng, X., Qiao, W., Li, L., Jiang, C., & Ni, L. (2020). Model of mining-induced fracture stress-seepage coupling in coal seam overburden and prediction of mine inflow. *Journal of China Coal Society, 45*(8), 2890–2900. (in Chinese).

Chi, M., Zhang, D., Honglin, L., Hongzhi, W., Yazhou, Z., Shuai, Z., & Qiang, Z. (2019). Simulation analysis of water resource damage feature and development degree of mining-induced fracture at ecologically fragile mining area. *Environmental Earth Sciences, 78*(3), 1–15.

Climent, N., Arroyo, M., O'Sullivan, C., & Gens, A. (2014). Sand production simulation coupling DEM with CFD. *European Journal of Environmental and Civil Engineering, 18*(9), 983–1008.

Cui, X., Li, J., Chan, A., & Chapman, D. (2014). Coupled DEM–LBM simulation of internal fluidisation induced by a leaking pipe. *Powder Technology, 254*, 299–306.

Cui, Y., Nouri, A., Chan, D., & Rahmati, E. (2016). A new approach to DEM simulation of sand production. *Journal of Petroleum Science and Engineering, 147*, 56–67.

Cundall, P. A., & Strack, O. D. (1979). A discrete numerical model for granular assemblies. *Geotechnique, 29*(1), 47–65.

Dehestani, A., Hosseini, M., & Beydokhti, A. T. (2020). Effect of wetting–drying cycles on mode I and mode II fracture toughness of sandstone in natural (pH= 7) and acidic (pH= 3) environments. *Theoretical and Applied Fracture Mechanics, 107*, 102512.

Deng, M. D., Fang, Z., & Liu, X. (1997). Research on the action of water in the inf rared radiation of the rocks. *Earthquake Research in China, 13*(3), 288–296.

Deng, P., Xu, K., & Guo, S. (2023). Effects of coarse aggregate morphology on concrete mechanical properties. *Journal of Building Engineering, 63*, 105408.

Di Felice, R. (1994). The voidage function for fluid-particle interaction systems. *International Journal of Multiphase Flow, 20*(1), 153–159.

Dong, Y., & Qiao, P. (2021). CT image-based synthetic mesostructure generation for multiscale fracture analysis of concrete. *Construction and Building Materials, 296*, 123582.

Du, B., Bai, H., & Wu, G. (2019a). Dynamic compression properties and deterioration of red-sandstone subject to cyclic wet-dry treatment. *Advances in Civil Engineering, 2019*, 1–10.

Du, J., Bu, Y., Shen, Z., & Cao, X. (2019b). Maximum penetration depth and penetration time predicting model of cementing fluid flow through wellbore into weakly consolidated formation. *Fractals, 27*(8), 1950132.

Du, F., Wang, K., Dong, X. L., & Wei, J. P. (2021). Numerical simulation of damage and failure of coal-rock combination based on CT three-dimensional reconstruction. *Journal of China Coal Society, 46*(S1), 253–262.

Du, F., Wang, K., Zhang, G., Zhang, Y., Zhang, G., & Wang, G. (2022a). Damage characteristics of coal under different loading modes based on CT three-dimensional reconstruction. *Fuel, 310*, 122304.

Du, J., Zhou, A., Shen, S. L., & Bu, Y. (2022b). Fractal-based model for maximum penetration distance of grout slurry flowing through soils with different dry densities. *Computers and Geotechnics, 141*, 104526.

Dudek, M., Tajduś, K., Misa, R., & Sroka, A. (2020). Predicting of land surface uplift caused by the flooding of underground coal mines–A case study. *International Journal of Rock Mechanics and Mining Sciences, 132*, 104377.

Eberhardt, E., Stimpson, B., & Stead, D. (1999). Effects of grain size on the initiation and propagation thresholds of stress-induced brittle fractures. *Rock Mechanics and Rock Engineering, 32*, 81–99.

Erguler, Z. A., & Ulusay, R. (2009). Water-induced variations in mechanical properties of clay-bearing rocks. *International Journal of Rock Mechanics and Mining Sciences, 46*(2), 355–370.

Erhunmwun, I. D., & Ikponmwosa, U. B. (2017). Review on finite element method. *Journal of Applied Sciences and Environmental Management, 21*(5), 999–1002.

Eslami, J., Grgic, D., & Hoxha, D. (2010). Estimation of the damage of a porous limestone from continuous (P-and S-) wave velocity measurements under uniaxial loading and different hydrous conditions. Geophysical Journal International, 183(3), 1362–1375.

Esterhuizen, G. S., & Karacan, C. O. (2005). Development of numerical models to investigate permeability changes and gas emission around longwall mining panel. In *The 40th US Symposium on Rock Mechanics (USRMS)*. American Rock Mechanics Association.

Esterhuizen, G., & Karacan, C. (2007). A methodology for determining gob permeability distributions and its application to reservoir modeling of coal mine longwalls. *SME Annual Conference, 88*, 012037.

Fan, L. M., Xiang, M. X., Peng, J., et al. (2016). Groundwater response to intensive mining in ecologically fragile area. *Journal of China Coal Society, 41*(11), 2672–2678. (in Chinese).

Fan, M., McClure, J., Han, Y., Li, Z., & Chen, C. (2018). Interaction between proppant compaction and single-/multiphase flows in a hydraulic fracture. *SPE Journal, 23*(4), 1290–1303.

Fan, J., Xie, H., Chen, J., Jiang, D., Li, C., Tiedeu, W. N., & Ambre, J. (2020). Preliminary feasibility analysis of a hybrid pumped-hydro energy storage system using abandoned coal mine goafs. *Applied Energy, 258*, 114007.

Fan, L. F., Fan, Y. D., Xi, Y., & Gao, J. W. (2022). Spatial failure mode analysis of frozen sandstone under uniaxial compression based on CT technology. *Rock Mechanics and Rock Engineering, 55*(7), 4123–4138.

Fang, J., Song, H., Xu, J., Yang, L., & Li, Z. (2019). Storage coefficient calculation model of coal mine underground reservoir considering effect of effective stress. *Journal of China Coal Society, 44*(12), 3750–3759. (in Chinese).

Farhidzadeh, A., Mpalaskas, A. C., Matikas, T. E., Farhidzadeh, H., & Aggelis, D. G. (2014). Fracture mode identification in cementitious materials using supervised pattern recognition of acoustic emission features. Construction and building materials 67, 129–138.

Feng, M., Wu, J., Ma, D., Ni, X., Yu, B., & Chen, Z. (2018b). Experimental investigation on the seepage property of saturated broken red sandstone of continuous gradation. *Bulletin of Engineering Geology and the Environment, 77*(3), 1167–1178.

Feng, Z., Feng, C., Zhong, Y., Qin, Z., Mao, R., Zhao, L., & Zong, X. (2022). TOC estimation of shale oil reservoir by combining nuclear magnetic resonance logging and nuclear physics logging. *Journal of Geophysics and Engineering, 19*(4), 833–845.

Feucht, L. J., & Logan, J. M. (1990). Effects of chemically active solutions on shearing behavior of a sandstone. *Tectonophysics, 175*(1–3), 159–176.

Fu, Y., Wang, Z., Liu, X., Yuan, W., Miao, L., Liu, J., & Dun, Z. (2017). Meso damage evolution characteristics and macro degradation of sandstone under wetting-drying cycles. *Chinese Journal of Geotechnical Engineering, 12*(4), 1653–1661. (in Chinese).

Fu, Y., Chen, X., & Feng, Z. L. (2020a). Characteristics of coal-rock fractures based on CT scanning and its influence on failure modes. *Journal of China Coal Society, 45*(2), 568–578.

Fu, Y., Feng, Z. L., & Li, P. C. (2020b). Heterogeneous characteristics of coal based on CT scanning and numerical analysis of its reconstruction model. *Journal of Mining & Safety Engineering, 37*(4), 828–835.

Fu, J., Labuz, J. F., Cheng, H., Hou, R., & Zhu, W. (2022). Simulating progressive failure in fractured saturated rock under seepage condition using a novel coupled model and the application. *Geomechanics and Geophysics for Geo-Energy and Geo-Resources, 8*(2), 1–19.

Fujii, Y., Saito, S., Oshima, T., Kodama, J. I., Fukuda, D., Sakata, S., & Dassanayake, A. B. (2020). Complete slaking collapse of dike sandstones by fresh water and prevention of the collapse by salt water. *International Journal of Rock Mechanics and Mining Sciences, 131*, 104378.

Fusi, N., & Martinez-Martinez, J. (2013). Mercury porosimetry as a tool for improving quality of micro-CT images in low porosity carbonate rocks. *Engineering Geology, 166*, 272–282.

Gao, F., & Stead, D. (2014). The application of a modified Voronoi logic to brittle fracture modelling at the laboratory and field scale. *International Journal of Rock Mechanics and Mining Sciences, 68*, 1–14.

Gao, X. G., Sha, J. F., Luan, J. Y., Li, M. C., & Wang, J. P. (2022). A triple-source CT system for micro-scale investigation of geological materials: A simulation study. *Applied Radiation and Isotopes, 190*, 110510.

Godinho, J. R., Grilo, B. L., Hellmuth, F., & Siddique, A. (2021). Mounted single particle characterization for 3D mineralogical analysis—MSPaCMAn. *Minerals, 11*(9), 947.

Godinho, J. R. A., Hassanzadeh, A., & Heinig, T. (2023). 3D quantitative mineral characterization of particles using X-ray computed tomography. *Natural Resources Research, 32*(2), 479–499.

Gong, C., Wang, W., Shao, J., Wang, R., & Feng, X. (2021). Effect of water chemical corrosion on mechanical properties and failure modes of pre-fissured sandstone under uniaxial compression. *Acta Geotechnica, 16*(4), 1083–1099.

Grasselli, G., & Egger, P. (2003). Constitutive law for the shear strength of rock joints based on three-dimensional surface parameters. *International Journal of Rock Mechanics and Mining Sciences, 40*(1), 25–40.

Gratchev, I., Pathiranagei, S. V., & Kim, D. H. (2019). Strength properties of fresh and weathered rocks subjected to wetting–drying cycles. *Geomechanics and Geophysics for Geo-Energy and Geo-Resources, 5*(3), 211–221.

Gu, D. (2013). Water resource and surface ecology protection technology of modern coal mining in China's energy "Golden Triangle". *Engineering and Science, 15*(4), 102–107.

Gu, D. (2015). Theory framework and technological system of coal mine underground reservoir. *Journal of China Coal Society, 40*(2), 239–246.

Gu, D., Li, J., Cao, Z., Wu, B., Jiang, B., Yang, Y., Yang, J., & Chen, Y. (2021). Technology and engineering development strategy of water protection and utilization of coal mine in China. *Journal of China Coal Society, 46*(10), 3079–3089. (in Chinese).

Guha Roy, D., Singh, T. N., Kodikara, J., & Das, R. (2017). Effect of water saturation on the fracture and mechanical properties of sedimentary rocks. *Rock Mechanics and Rock Engineering, 50*(10), 2585–2600.

Guntoro, P. I., Ghorbani, Y., Parian, M., Butcher, A. R., Kuva, J., & Rosenkranz, J. (2021). Development and experimental validation of a texture-based 3D liberation model. *Minerals Engineering, 164*, 106828.

Guo, J., Feng, G. R., Qi, T., Wang, P., Yang, J., Li, Z., & Wang, Z. (2018). Dynamic mechanical behavior of dry and water saturated igneous rock with acoustic emission monitoring. *Shock and Vibration, 2018*, 2348394.

Guo, P., Gu, J., Su, Y., Wang, J., & Ding, Z. (2021). Effect of cyclic wetting–drying on tensile mechanical behavior and microstructure of clay-bearing sandstone. *International Journal of Coal Science & Technology, 8*(5), 956–968.

Gupta, N., Mishra, B., & Crandall, D. M. (2022). A new workflow of X-ray CT image processing and data analysis of structural features in rock using open-source software. *Mining, Metallurgy & Exploration, 39*(5), 2011–2024.

Guyer, J. E., Wheeler, D., & Warren, J. A. (2009). FiPy: Partial differential equations with Python. *Computing in Science & Engineering, 11*(3), 6–15.

Haberfield, C. M., & Johnston, I. W. (1990). Determination of the fracture toughness of a saturated soft rock. *Canadian Geotechnical Journal, 27*(3), 276–284.

Han, Y., & Cundall, P. A. (2011). Lattice Boltzmann modeling of pore-scale fluid flow through idealized porous media. *International Journal for Numerical Methods in Fluids, 67*(11), 1720–1734.

Han, P., Zhang, C., & Wang, W. (2022a). Failure analysis of coal pillars and gateroads in longwall faces under the mining-water invasion coupling effect. *Engineering Failure Analysis, 131*, 105912.

Han, P., Zhang, C., He, X., Wang, X., & Qiao, Y. (2022b). DEM fluid–solid coupling method for progressive failure simulation of roadways in a fault structure area with water-rich roofs. *Geomechanics and Geophysics for Geo-Energy and Geo-Resources, 8*(6), 194.

Han, P., Zhang, C., Wang, X., & Wang, L. (2023). Study of mechanical characteristics and damage mechanism of sandstone under long-term immersion. *Engineering Geology, 315*, 107020.

Hao, Y., & Azzam, R. (2005). The plastic zones and displacements around underground openings in rock masses containing a fault. *Tunnelling and Underground Space Technology, 20*(1), 49–61.

Hao, C. G., Guo, X. Y., Deng, C. B., Zhang, X. H., Zhao, B., & Wang, X. (2022). Precise identification and threshold inversion of pores and fissures in CT digital coal rock based on Bi-PTI model. *Journal of China Coal Society, 48*, 1–11.

Hashiba, K., & Fukui, K. (2015). Effect of water on the deformation and failure of rock in uniaxial tension. *Rock Mechanics and Rock Engineering, 48*(5), 1751–1761.

Hawkins, A. B., & McConnell, B. J. (1992). Sensitivity of sandstone strength and deformability to changes in moisture content. Quarterly Journal of Engineering Geology and Hydrogeology, 25(2), 115–130.

He, M., Zhou, L., Li, D., Wang, C., & Nie, W. (2008). Experimental research on hydrophilic characteristics of mudstone in deep well. *Chinese Journal of Rock Mechanics and Engineering, 27*(6), 1113–1120. (in Chinese).

He, J. Q., Wei, H. Z., Meng, Q. S., Wang, X. Z., & Wei, C. F. (2018). Evolution of particle breakage of calcareous sand under large displacement shearing. *Rock and Soil Mechanics, 39*(1), 165–172.

He, L., Yu, J., Hu, Q., Cai, Q., Qu, M., & He, T. (2020a). Study on crack propagation and shear behavior of weak muddy intercalations submitted to wetting-drying cycles. *Bulletin of Engineering Geology and the Environment, 79*(9), 4873–4889.

He, X., Zhao, Y., Zhang, C., & Han, P. (2020b). A model to estimate the height of the water-conducting fracture zone for longwall panels in Western China. *Mine Water and the Environment, 39*(4), 823–838.

Heggheim, T., Madland, M. V., & Risnes, R. (2005). A chemical induced enhanced weakening of chalk by seawater. *Journal of Petroleum Science and Engineering, 46*(3), 171–184.

Heyn, R., Rozendaal, A., Du Plessis, A., & Mouton, C. (2021). Characterization of coloured gemstones by X-ray micro computed tomography. *Minerals, 11*(2), 178.

Holmes, D. W., Williams, J. R., & Tilke, P. (2011). Smooth particle hydrodynamics simulations of low Reynolds number flows through porous media. *International Journal for Numerical and Analytical Methods in Geomechanics, 35*(4), 419–437.

Hou, P., Liang, X., Zhang, Y., He, J., Gao, F., & Liu, J. (2021). 3D multi-scale reconstruction of fractured shale and influence of fracture morphology on shale gas flow. *Natural Resources Research, 30*, 2463–2481.

Hu, H., Zhang, W., Xing, Y., & Li, X. (2017). Molecular Dynamics Study on the Effect of Single Grain Boundary on the Mechanical Behavior of α-Quartz. International Journal of Rock Mechanics and Mining Sciences, 93, 94–100.

Hu, Y., Sun, J., Liu, W., & Wei, D. (2019). The evolution and prevention of water inrush due to fault activation at working face No. II 632 in the Hengyuan coal mine. *Mine Water and the Environment, 38*(1), 93–103.

Hua, W., Dong, S., Li, Y., Xu, J., & Wang, Q. (2015). The influence of cyclic wetting and drying on the fracture toughness of sandstone. *International Journal of Rock Mechanics and Mining Sciences, 100*(78), 331–335.

Huang, S., Xia, K., Yan, F., & Feng, X. (2010). An experimental study of the rate dependence of tensile strength softening of Longyou sandstone. *Rock Mechanics and Rock Engineering, 43*(6), 677–683.

Huang, Z., Jiang, Z., Zhu, S., Wu, X., Yang, L., & Guan, Y. (2016). Influence of structure and water pressure on the hydraulic conductivity of the rock mass around underground excavations. *Engineering Geology, 202*, 74–84.

Huang, S., Wang, J., Qiu, Z., & Kang, K. (2018). Effects of cyclic wetting-drying conditions on elastic modulus and compressive strength of sandstone and mudstone. *Processes, 6*(12), 234.

Huang, X., Pang, J., Liu, G., & Chen, Y. (2020a). Experimental study on physicomechanical properties of deep sandstone by coupling of dry-wet cycles and acidic environment. *Advances in Civil Engineering, 2020*, 2760952.

Huang, X., Pang, J., Zou, J. (2022). Study on the Effect of Dry–Wet Cycles on Dynamic Mechanical Properties of Sandstone Under Sulfuric Acid Solution. *Rock Mechanics and Rock Engineering*. 1–17.

Huang, Z., Zuo, Q., Wu, L., Chen, F., Hu, S., & Zhu, S. (2020b). Nonlinear softening mechanism of argillaceous slate under water-rock interaction. *Rock and Soil Mechanics, 41*(9), 2931–2942. (in Chinese).

Huillca, Y., Silva, M., Ovalle, C., Quezada, J. C., Carrasco, S., & Villavicencio, G. E. (2021). Modelling size effect on rock aggregates strength using a DEM bonded-cell model. *Acta Geotechnica, 16*(3), 699–709.

Huo, D., & Benson, S. M. (2016). Experimental investigation of stress-dependency of relative permeability in rock fractures. *Transport in Porous Media, 113*(3), 567–590.

Huo, D., Li, B., & Benson, S. M. (2014). Investigating aperture-based stress-dependent permeability and capillary pressure in rock fractures. In *SPE annual technical conference and exhibition*. OnePetro.

Hustrulid, W. A. (1976). A review of coal pillar strength formulas. *Rock Mechanics and Rock Engineering, 8*(2), 115–145.

Itasca Consulting Group. (2019). FLAC3D Version 6.0 documentation.

Jiang, J., Zhang, Z., Wang, D., Wang, L., & Han, X. (2022a). Web pillar stability in open-pit highwall mining. *International Journal of Coal Science & Technology, 9*(1), 1–16.

Jiang, M. J., & Zhang, W. C. (2014). Coupled CFD-DEM method for soils incorporating equation of state for liquid. Yantu Gongcheng Xuebao/Chinese Journal of Geotechnical Engineering, 36(5), 793–801.

Jiang, M. J., Yu, H. S., & Harris, D. (2005). A novel discrete model for granular material incorporating rolling resistance. *Computers and Geotechnics*, 32(5), 340-357.

Jiang, Q., Cui, J., Feng, X., & Jiang, Y. (2014). Application of computerized tomographic scanning to the study of water-induced weakening of mudstone. Bulletin of Engineering Geology and the Environment, 73(4), 1293–1301.

Jiang, X., Li, C., Zhou, J. Q., Zhang, Z., Yao, W., Chen, W., & Liu, H. B. (2022b). Salt-induced structure damage and permeability enhancement of Three Gorges Reservoir sandstone under wetting-drying cycles. *International Journal of Rock Mechanics and Mining Sciences, 153*, 105100.

Kamel, K. E. M., Gerard, P., Colliat, J. B., & Massart, T. J. (2022). Modelling stress-induced permeability alterations in sandstones using CT scan-based representations of the pore space morphology. *International Journal of Rock Mechanics and Mining Sciences, 150*, 104998.

Kataoka, M., Obara, Y., & Kuruppu, M. (2015). Estimation of fracture toughness of anisotropic rocks by semi-circular bend (SCB) tests under water vapor pressure. *Rock Mechanics and Rock Engineering, 48*(4), 1353–1367.

Khanlari, G., & Abdilor, Y. (2015). Influence of wet–dry, freeze–thaw, and heat–cool cycles on the physical and mechanical properties of upper red sandstones in Central Iran. *Bulletin of Engineering Geology and the Environment, 74*(4), 1287–1300.

Kim, D. H., Gratchev, I., & Balasubramaniam, A. (2013). Determination of joint roughness coefficient (JRC) for slope stability analysis: A case study from the Gold Coast area, Australia. *Landslides, 10*(5), 657–664.

Kloss, C., Goniva, C., Hager, A., Amberger, S., & Pirker, S. (2012). Models, algorithms and validation for opensource DEM and CFD–DEM. *Progress in Computational Fluid Dynamics, an International Journal, 12*(2–3), 140–152.

Kong, H., & Wang, L. (2018). Seepage problems on fractured rock accompanying with mass loss during excavation in coal mines with karst collapse columns. *Arabian Journal of Geosciences, 11*(19), 1–13.

Kong, X., Wang, E., He, X., Li, Z., Li, D., & Liu, Q. (2017). Multifractal characteristics and acoustic emission of coal with joints under uniaxial loading. *Fractals-Complex Geometry, Patterns, and Scaling in Nature and Society, 25*(5), 1750045.

Kong, X., Wang, E., Li, S., Lin, H., Xiao, P., Zhang, K. (2019). Fractals and chaos characteristics of acoustic emission energy about gasbearing coal during loaded failure. Fractals-Complex Geom. Patterns Scaling Nat. Soc. 27(05), 1950072.

Koponen, A., Kataja, M., & Timonen, J. (1997). Permeability and effective porosity of porous media. *Physical Review E, 56*(3), 3319–3325.

Kruggel-Emden, H., Kravets, B., Suryanarayana, M. K., & Jasevicius, R. (2016). Direct numerical simulation of coupled fluid flow and heat transfer for single particles and particle packings by a LBM-approach. *Powder Technology, 294*, 236–251.

Kuang, N. J., Zhou, J. P., Xian, X. F., Zhang, C. P., Yang, K., & Dong, Z. Q. (2023). Geomechanical risk and mechanism analysis of CO2 sequestration in unconventional coal seams and shale gas reservoirs. *Rock Mechanics Bulletin, 2*(4), 100079.

Kubeyev, A., Forbes Inskip, N., Phillips, T., Zhang, Y., Maier, C., Bisdom, K., & Doster, F. (2022). Digital image-based stress–permeability relationships of rough fractures using numerical contact mechanics and stokes equation. *Transport in Porous Media, 141*, 1–36.

Li, G., & Liang, B. (2009). Experimental research on the effect of pore water pressure on the creep laws of soft rock. *Journal of China Coal Society, 34*(8), 1067–1070. (in Chinese).

Li, B., & Wu, Q. (2019). Catastrophic evolution of water inrush from a water-rich fault in front of roadway development: A case study of the hongcai coal mine. *Mine Water and the Environment, 38*(2), 421–430.

Li, J. B., & Xu, Y. C. (2016). Mechanical model of the collapse column water inrush prevention considering the confined water seepage and its application. *Journal of China University of Mining & Technology, 45*(2), 217–224.

Li, L., Tang, C., Zuo, Y., Li, G., & Liu, C. (2009). Mechanism of hysteretic groundwater inrush from coal seam floor with karstic collapse columns. *Journal of China Coal Society, 34*(9), 1212–1216. (in Chinese).

Li, D., Wang, G., Han, L., Liu, P., He, M., Yang, G., & Chen, C. (2011). Analysis of microscopic pore structures of rocks before and after water absorption. *Mining Science and Technology, 21*(2), 287–293.

Li, Z., Xie, H., Li, J., & He, Z. (2014). Experimental study of mining effect on collapse column activated water conducting mechanism. *Journal of Central South University (Science and Technology), 45*(12), 4377–4383. (in Chinese).

Li, G., Zhang, R., Xu, X. L., & Zhang, Y. F. (2015). CT image reconstruction of coal rock three-dimensional fractures and body fractal dimension under triaxial compression test. *Rock and Soil Mechanics, 36*(6), 1633–1642.

Li, H., Shen, R., Li, D., Jia, H., Li, T., Chen, T., & Hou, Z. (2019a). Acoustic emission multi-parameter analysis of dry and saturated sandstone with cracks under uniaxial compression. *Energies, 12*(10), 1959.

Li, S., Gao, C., Zhou, Z., Li, L., Wang, M., Yuan, Y., & Wang, J. (2019b). Analysis on the precursor information of water inrush in karst tunnels: A true triaxial model test study. *Rock Mechanics and Rock Engineering, 52*(2), 373–384.

Li, H., Shen, R., Wang, E., Li, D., Li, T., Chen, T., & Hou, Z. (2020a). Effect of water on the time-frequency characteristics of electromagnetic radiation during sandstone deformation and fracturing. *Engineering Geology, 265*, 105451. (in Chinese).

Li, J., Huang, Y., Gao, H., Ouyang, S., & Guo, Y. (2020b). Transparent characterization of spatial-temporal evolution of gangue solid wastes' void structures during compression based on CT scanning. *Powder Technology, 376*, 477–485.

Li, Y., Cui, H., Zhang, P., Wang, D., & Wei, J. (2020c). Three-dimensional visualization and quantitative characterization of coal fracture dynamic evolution under uniaxial and triaxial compression based on μCT scanning. *Fuel, 262*, 116568.

Li, Z., Xiong, Z., Chen, H., Lu, H., Huang, M., Ma, C., & Liu, Y. (2020d). Analysis of stress-strain relationship of brittle rock containing microcracks under water pressure. *Bulletin of Engineering Geology and the Environment, 79*, 1909–1918.

Li, S., Zhang, Z., & Yao, X. (2021). Cracking process and energy dissipation of sandstone under repetitive impact loading with different loading rates: From micro to macro scale. Construction and Building Materials, 302, 124123.

Li, E., Feng, J., Zhang, L., Zhang, H., & Zhu, T. (2021a). Brazilian tests on layered carbonaceous slate under water-rock interaction and weathering. *Chinese Journal of Geotechnical Engineering, 43*(2), 329–337. (in Chinese).

Li, H., Qiao, Y., Shen, R., He, M., Cheng, T., Xiao, Y., & Tang, J. (2021b). Effect of water on mechanical behavior and acoustic emission response of sandstone during loading process: Phenomenon and mechanism. *Engineering Geology, 294*, 106386.

Li, X., Peng, K., Peng, J., & Hou, D. (2021c). Experimental investigation of cyclic wetting-drying effect on mechanical behavior of a medium-grained sandstone. *Engineering Geology, 293*, 106335.

Li, X., Cai, J., Chen, D., & Feng, Q. (2021d). Characteristics of water contamination in abandoned coal mines: A case study on Yudong River area, Kaili, Guizhou Province, China. *International Journal of Coal Science & Technology, 8*(6), 1491–1503.

Li, B., Zhang, G., Ma, W., Liu, M., & Li, A. (2022a). Damage mechanism of sandstones subject to cyclic freeze–thaw (FT) actions based on high-resolution computed tomography (CT). *Bulletin of Engineering Geology and the Environment, 81*(9), 374.

Li, Y. B., Zhai, Y., Li, Y., Meng, F. D., & Zhang, Y. S. (2022b). Sandstone impact mechanical properties and numerical simulation after freeze-thaw cycles. *Chinese Journal of Underground Space and Engineering, 18*(5), 1580–1587.

Li, H., Cao, Z., Wu, B., Chi, M., & Zhang, B. (2023). Experimental study on characteristics of grounawater fracture in coalmine overlying rock. *Coal Science and Technology, 51*(6), 168–176.

Liang, H., & Fu, Y. (2020). Fracture properties of sandstone degradation under the action of drying–wetting cycles in acid and alkaline environments. *Arabian Journal of Geosciences, 13*(2), 1–8.

Lin, Q., Cao, P., Cao, R., & Fan, X. (2019). Acoustic emission characteristics during rock fragmentation processes induced by disc cutter under different water content conditions. *Applied Sciences, 9*(1), 194.

Liu, W., & Zhang, Z. (2020). Experimental characterization and quantitative evaluation of slaking for strongly weathered mudstone underw cyclic wetting-drying condition. *Arabian Journal of Geosciences, 13*(2), 1–8.

Liu, S., Wu, L., Zhang, Y., & Chen, Q. (2010). Change feature of infrared radiation from loaded damp rock. *Journal of Northeastern University, 31*(2), 1034–1038.

Liu, X., Fu, Y., Zheng, Y., & Liang, H. (2012). A review on deterioration of rock caused by water-rock interaction. *Chinese Journal of Underground Space and Engineering, 8*(1), 77–82. (in Chinese).

Liu, X., Wang, Z., Fu, Y., Yuan, W., & Miao, L. (2016). Macro/microtesting and damage and degradation of sandstones under dry-wet cycles. *Advances in Materials Science and Engineering, 2016*, 7013032.

Liu, J., Chen, W., Liu, T., Yu, J., Dong, J., & Nie, W. (2018a). Effects of initial porosity and water pressure on seepage-erosion properties of water inrush in completely weathered granite. *Geofluids, 2018*, 4103645.

Liu, X., Jin, M., Li, D., & Zhang, L. (2018b). Strength deterioration of a shaly sandstone under dry–wet cycles: A case study from the Three Gorges Reservoir in China. *Bulletin of Engineering Geology and the Environment, 77*(4), 1607–1621.

Liu, X., Wu, L., Zhang, Y., Liang, Z., Yao, X., & Liang, P. (2019). Frequency properties of acoustic emissions from the dry and saturated rock. *Environment and Earth Science, 78*(3), 1–17.

Liu, Y., Yin, G., Li, M., Zhang, D., Huang, G., Liu, P., & Yu, B. (2020). Mechanical properties and failure behavior of dry and water-saturated anisotropic coal under true-triaxial loading conditions. *Rock Mechanics and Rock Engineering, 53*, 4799–4818.

Liu, C., Cheng, Y., Jiao, Y., Zhang, G., Zhang, W., Ou, G., & Tan, F. (2021). Experimental study on the effect of water on mechanical properties of swelling mudstone. *Engineering Geology, 295*, 106448.

Liu, J., Zhao, Y., Tan, T., Zhang, L., Zhu, S., & Xu, F. (2022). Evolution and modeling of mine water inflow and hazard characteristics in southern coalfields of China: A case of Meitanba mine. *International Journal of Mining Science and Technology, 32*(3), 513–524.

Lu, G., Crandall, D., & Bunger, A. P. (2021). Observations of breakage for transversely isotropic shale using acoustic emission and X-ray computed tomography: Effect of bedding orientation, pre-existing weaknesses, and pore water. *International Journal of Rock Mechanics and Mining Sciences, 139*, 104650.

Luo, Y. (2020). Influence of water on mechanical behavior of surrounding rock in hard-rock tunnels: An experimental simulation. *Engineering Geology, 277*, 105816.

Luo, J., Tang, H., & Sui, Z. (2021). Hydrochemistry of coal samples in water immersion process. *Science Technology and Engineering, 21*(29), 12431–12437.

Luo, X., Zhang, Y., Zhou, H., He, K., Zhang, B., Zhang, D., & Xiao, W. (2022). Pore structure characterization and seepage analysis of ionic rare earth orebodies based on computed tomography images. *International Journal of Mining Science and Technology, 32*(2), 411–421.

Ma, D., Miao, X., Chen, Z., & Mao, X. (2013). Experimental investigation of seepage properties of fractured rocks under different confining pressures. *Rock Mechanics and Rock Engineering, 46*(5), 1135–1144.

Ma, D., Rezania, M., Yu, H., & Bai, H. (2017). Variations of hydraulic properties of granular sandstones during water inrush: Effect of small particle migration. *Engineering Geology, 217*, 61–70.

Ma, Q., Yu, P., & Yuan, P. (2018). Experimental study on the influence of dry wet cycle on creep characteristics of deep siltstone. *Chinese Journal of Rock Mechanics and Engineering, 37*(3), 593–600.

Ma, D., Duan, H., Li, X., Li, Z., Zhou, Z., & Li, T. (2019). Effects of seepage-induced erosion on nonlinear hydraulic properties of broken red sandstones. *Tunnelling and Underground Space Technology, 91*, 102993.

Ma, D., Duan, H., Zhang, J., Feng, X., & Huang, Y. (2021). Experimental investigation of creep-erosion coupling mechanical properties of water inrush hazards in fault fracture rock masses. *Chinese Journal of Rock Mechanics and Engineering, 40*(9), 1751–1763. (in Chinese).

Ma, D., Duan, H., Zhang, J., & Bai, H. (2022a). A state-of-the-art review on rock seepage mechanism of water inrush disaster in coal mines. *International Journal of Coal Science & Technology, 9*(1), 1–28.

Ma, Y. L., Wang, C. R., & Ma, K. (2022b). SEM and permeability enhancement experiment study on thermal damage characteristics of coal-rock under infrared radiation. *Coal Science and Technology, 50*(7), 177–183.

Maerz, N. H., Franklin, J. A., & Bennett, C. P. (1990). Joint roughness measurement using shadow profilometry. *International Journal of Rock Mechanics and Mining Sciences & Geomechanics Abstracts, 27*(5), 329–343.

Mao, Y. J., Chen, X., Fan, C. N., Ge, S. C., & Li, W. P. (2022). Crack network evolution of water injection coal and rock mass by means of 3D reconstruction. *Rock and Soil Mechanics, 43*(6), 1717–1726.

Maruvanchery, V., & Kim, E. (2019). Effects of water on rock fracture properties: Studies of mode I fracture toughness, crack propagation velocity, and consumed energy in calcite-cemented sandstone. *Geomechanics and Engineering, 17*(1), 57–67.

Masoumi, H., Horne, J., & Timms, W. (2017). Establishing empirical relationships for the effects of water content on the mechanical behavior of Gosford sandstone. Rock Mechanics and Rock Engineering, 50(8), 2235–2242.

Menéndez, J., Ordóñez, A., Álvarez, R., & Loredo, J. (2019). Energy from closed mines: Underground energy storage and geothermal applications. *Renewable and Sustainable Energy Reviews, 108*, 498–512.

Meng, F., Song, J., Wong, L. N. Y., Wang, Z., & Zhang, C. (2021). Characterization of roughness and shear behavior of thermally treated granite fractures. Engineering Geology. 293, 106287.

Meng, X., Qin, Y., Zhuge, F., & Liu, W. (2022). Microstructure characteristics and the damage evolution rule of sandstone under triaxial compression test using Nuclear Magnetic Resonance (NMR). Materials Letters. 309, 131355.

Meng, Z., Zhang, J., & Wang, R. (2011). In-situ stress, pore pressure and stress-dependent permeability in the Southern Qinshui Basin. *International Journal of Rock Mechanics and Mining Sciences, 48*(1), 122–131.

Meng, Z., Shi, X., & Li, G. (2016). Deformation, failure and permeability of coal-bearing strata during longwall mining. *Engineering Geology, 208*, 69–80.

Miao, C., Yang, L., Xu, Z., Yang, K., Sun, X., Jiang, M., & Zhao, W. (2021). Experimental study on strength softening behaviors and micro-mechanisms of sandstone based on nuclear magnetic resonance. *Chinese Journal of Rock Mechanics and Engineering, 40*(11), 2189–2198.

Milne, D., Hawkes, C., & Hamilton, C. (2009, May). A new tool for the field characterization of joint surfaces. In *Proceedings of the 3rd CANUS rock mechanics symposium.*

Moffat, R., & Fannin, R. J. (2011). A hydromechanical relation governing internal stability of cohesionless soil. *Canadian Geotechnical Journal, 48*(3), 413–424.

Momeni, A., Hashemi, S. S., Khanlari, G. R., & Heidari, M. (2017). The effect of weathering on durability and deformability properties of granitoid rocks. *Bulletin of Engineering Geology and the Environment, 76*(3), 1037–1049.

Moon, J., & Fernandez, G. (2010). Effect of excavation-induced groundwater level drawdown on tunnel inflow in a jointed rock mass. *Engineering Geology, 110*(3–4), 33–42.

Nara, Y., Tanaka, M., & Harui, T. (2017). Evaluating long-term strength of rock under changing environments from air to water. Engineering Fracture Mechanics. 178, 201–211.

National Bureau of Statistics. (2021). *China statistical yearbook-2021*. China Statistics Press.

Nguyen, C. D., Benahmed, N., Andò, E., Sibille, L., & Philippe, P. (2019). Experimental investigation of microstructural changes in soils eroded by suffusion using X-ray tomography. *Acta Geotechnica, 14*(3), 749–765.

Ni, X., Niu, Y., Wang, Y., & Yu, K. (2018). Non-Darcy flow experiments of water seepage through rough-walled rock fractures. *Geofluids, 2018*, 8541421.

Noël, C., Baud, P., & Violay, M. (2021). Effect of water on sandstone's fracture toughness and frictional parameters: Brittle strength constraints. *International Journal of Rock Mechanics and Mining Sciences, 147*, 104916.

Ohtsu, M. (1991). Simplified moment tensor analysis and unified decomposition of acoustic emission source: application to in situ hydrofracturing test. Journal of Geophysical Research: Solid Earth, 96(B4), 6211–6221.

Özbek, A. (2014). Investigation of the effects of wetting–drying and freezing–thawing cycles on some physical and mechanical properties of selected ignimbrites. *Bulletin of Engineering Geology and the Environment, 73*(2), 595–609.

Pan, Y., Wu, G., Zhao, Z., & He, L. (2020). Analysis of rock slope stability under rainfall conditions considering the water-induced weakening of rock. *Computers and Geotechnics, 128*, 103806.

Pang, Y., Li, Q., Cao, G., & Zhou, B. (2019). Analysis and calculation method of underground reservoir water storage space composition. *Journal of China Coal Society, 44*(2), 557–566. (in Chinese).

Peng, S. (2013). *Coal mine ground control* (3rd ed.). China University of Mining and Technology Press.

Peng, S., & Chiang, H. (1984). *Longwall mining*. Wiley.

Poulsen, B. A., Shen, B., Williams, D. J., Huddlestone-Holmes, C., Erarslan, N., & Qin, J. (2014). Strength reduction on saturation of coal and coal measures rocks with implications for coal pillar strength. *International Journal of Rock Mechanics and Mining Sciences, 71*, 41–52.

Pujades, E., Willems, T., Bodeux, S., Orban, P., & Dassargues, A. (2016). Underground pumped storage hydroelectricity using abandoned works (deep mines or open pits) and the impact on groundwater flow. *Hydrogeology Journal, 24*(6), 1531–1546.

Qian, M., Shi, P., & Xu, J. (2010). *Coal mine groundpressure and ground control*. China University of Mining and Technology.

Qiao, W., Li, W., Li, T., Chang, J., & Wang, Q. (2017). Effects of coal mining on shallow water resources in semiarid regions: A case study in the Shennan mining area, Shaanxi, China. *Mine Water and the Environment, 36*(1), 104–113.

Qiao, Y., Zhang, C., Zhang, L., & Zhao, W. (2020). Numerical simulation of fluid-solid coupling of fractured rock mass considering changes in fracture stiffness. *Energy Science & Engineering, 8*(1), 28–37.

Qin, Z., Chen, X., & Fu, H. (2018). Damage features of altered rock subjected to drying-wetting cycles. *Advances in Civil Engineering, 2018*, 5170832.

Qin, Y., Su, W., Tian, F., & Chen, Y. (2020). Research status and development direction of microcosmic effect under coal seam water injection. *Journal of China University of Mining and Technology, 49*(3), 428–444. (in Chinese).

Rabat, Á., Cano, M., Tomás, R., Tamayo, Á. E., & Alejano, L. R. (2020). Evaluation of strength and deformability of soft sedimentary rocks in dry and saturated conditions through needle penetration and point load tests: A comparative study. *Rock Mechanics and Rock Engineering, 53*(6), 2707–2723.

Renani, H. R., & Martin, C. D. (2018). Modeling the progressive failure of hard rock pillars. *Tunnelling and Underground Space Technology, 74*, 71–81.

Ramandi, H. L., Mostaghimi, P., Armstrong, R. T., Saadatfar, M., & Pinczewski, W. V. (2016). Porosity and permeability characterization of coal: a micro-computed tomography study. International Journal of Coal Geology, 154, 57–68.

Ranjith, P. G., Jasinge, D., Song, J., & Choi, S. K. (2008). A study of the effect of displacement rate and moisture content on the mechanical properties of concrete: Use of acoustic emission. *Mechanics of Materials, 40*(6), 453–469.

Ranjith, P. G., Perera, M. S. A., Perera, W. K. G., Choi, S. K., & Yasar, E. (2014). Sand production during the extrusion of hydrocarbons from geological formations: A review. *Journal of Petroleum Science and Engineering, 124*, 72–82.

Reviron, N., Reuschlé, T., & Bernard, J. D. (2009). The brittle deformation regime of water-saturated siliceous sandstones. Geophysical Journal International, 178(3), 1766–1778.

Richards, K. S., & Reddy, K. R. (2007). Critical appraisal of piping phenomena in earth dams. *Bulletin of Engineering Geology and the Environment, 66*(4), 381–402.

Rutqvist, J. (2011). Status of the TOUGH-FLAC simulator and recent applications related to coupled fluid flow and crustal deformations. *Computers & Geosciences, 37*, 739–750.

Sabat, L., & Kundu, C. K. (2021). History of finite element method: A review. In B. Das, S. Barbhuiya, R. Gupta, & P. Saha (Eds.), *Recent developments in sustainable infrastructure* (pp. 395–404). Springer.

Sadeghiamirshahidi, M., & Vitton, S. J. (2019). Laboratory study of gypsum dissolution rates for an abandoned underground mine. *Rock Mechanics and Rock Engineering, 52*(7), 2053–2066.

Salmi, E. F., Nazem, M., & Karakus, M. (2017). The effect of rock mass gradual deterioration on the mechanism of post-mining subsidence over shallow abandoned coal mines. *International Journal of Rock Mechanics and Mining Sciences, 91*, 59–71.

Salomon, E., Rotevatn, A., Kristensen, T. B., Grundvåg, S.-A., & Henstra, G. A. (2021). Microstructure and fluid flow in the vicinity of basin bounding faults in rifts – The Dombjerg fault, NE Greenland rift system. *Journal of Structural Geology, 153*, 104463.

Shao, X. Q., Chi, S. C., Tao, Y., & Zhou, X. X. (2020). DEM simulation of the size effect on the wetting deformation of rockfill materials based on single-particle crushing tests. *Computers and Geotechnics, 123*, 103429.

Shen, R., Li, H., Wang, E., Chen, T., Li, T., Tian, H., & Hou, Z. (2020). Infrared radiation characteristics and fracture precursor information extraction of loaded sandstone samples with varying moisture contents. *International Journal of Rock Mechanics and Mining Sciences, 130*, 104344.

Shi, X., Pan, J., Hou, Q., Jin, Y., Wang, Z., Niu, Q., & Li, M. (2018). Micrometer-scale fractures in coal related to coal rank based on micro-CT scanning and fractal theory. *Fuel, 212*, 162–172.

Shimizu, Y. (2011). Microscopic numerical model of fluid flow in granular material. *Geotechnique, 61*(10), 887–896.

Showkati, A., Salari-rad, H., & Aghchai, M. H. (2021). Predicting long-term stability of tunnels considering rock mass weathering and deterioration of primary support. *Tunnelling and Underground Space Technology, 107*, 103670.

Singh, K. K., Singh, D. N., & Ranjith, P. G. (2015). Laboratory simulation of flow through single fractured granite. *Rock Mechanics and Rock Engineering, 48*(3), 987–1000.

Song, W., & Liang, Z. (2021). Theoretical and numerical investigations on mining-induced fault activation and groundwater outburst of coal seam floor. *Bulletin of Engineering Geology and the Environment, 80*(7), 5757–5768.

Song, H., Jiang, Y., Elsworth, D., Zhao, Y., Wang, J., & Liu, B. (2018). Scale effects and strength anisotropy in coal. *International Journal of Coal Geology, 195*, 37–46.

Song, C., Ji, H., Liu, Z., Zhang, Y., Wang, H., & Tan, J. (2019). Experimental study on acoustic emission characteristics of weakly cemented rock under dry wet cycle. *Journal of Mining & Safety Engineering, 36*(4), 812–819.

Song, H., Zhao, Y., Jiang, Y., & Du, W. (2020). Experimental investigation on the tensile strength of coal: Consideration of the specimen size and water content. *Energies, 13*(24), 6585.

Sun, G., Wang, Z., Liu, P., & Zhang, Z. (2022). Dissolved micro-pores in alkali feldspar and their contribution to improved properties of tight sandstone reservoirs. Oil Gas Geol, 43(3), 658–669.

Sun, Q., Zhang, Y., 2019. Combined effects of salt, cyclic wetting and drying cycles on the physical and mechanical properties of sandstone. Engineering Geology. 248, 70–79.

Sun, X., Xu, H., Zheng, L., He, M., & Gong, W. (2016). An experimental investigation on acoustic emission characteristics of sandstone rockburst with different moisture contents. *Science China Technological Sciences, 59*(10), 1549–1558.

Sun, W., Xue, Y., Li, T., & Liu, W. (2019). Multi-field coupling of water inrush channel formation in a deep mine with a buried fault. *Mine Water and the Environment, 38*(3), 528–535.

Sun, H., Ma, L., Fu, Y., Han, J., Liu, S., Chen, M., & Tian, F. (2021a). Infrared radiation test on the influence of water content on sandstone damage evolution. *Infrared Physics & Technology, 118*, 103876.

Sun, Q., Ma, F., Guo, J., Zhao, H., Li, G., Liu, S., & Duan, X. (2021b). Excavation-induced deformation and damage evolution of deep tunnels based on a realistic stress path. *Computers and Geotechnics, 129*, 103843.

Sun, Q., Zhang, Y. (2019). Combined effects of salt, cyclic wetting and drying cycles on the physical and mechanical properties of sandstone. Engineering geology. 248, 70–79.

Sun, X., Miao, C., Jiang, M., Zhang, Y., Yang, L., & Guo, B. (2021c). Experimental and theoretical study on creep of sandstone with different moisture content based on modified Nishihara model. *Chinese Journal of Rock Mechanics and Engineering, 40*(12), 2411–2420.

Talesnick, M., & Shehadeh, S. (2007). The effect of water content on the mechanical response of a high-porosity chalk. *International Journal of Rock Mechanics and Mining Sciences, 44*(4), 584–600.

Tang, C., Yao, Q., Chen, T., Shan, C., & Li, J. (2022). Effects of water content on mechanical failure behaviors of coal samples. Geomechanics and Geophysics for Geo-Energy and Geo-Resources, 8(3), 1–17.

Tang, S. (2018). The effects of water on the strength of black sandstone in a brittle regime. *Engineering Geology, 239*, 167–178.

Tang, S. B., Yu, C. Y., Heap, M. J., Chen, P. Z., & Ren, Y. G. (2018). The influence of water saturation on the short-and long-term mechanical behavior of red sandstone. *Rock Mechanics and Rock Engineering, 51*(9), 2669–2687.

Tang, C., Yao, Q., Xu, Q., Shan, C., Xu, J., Han, H., & Guo, H. (2021a). Mechanical failure modes and fractal characteristics of coal samples under repeated drying–saturation conditions. *Natural Resources Research, 30*(6), 4439–4456.

Tang, D. X., Liu, W. L., Lou, Y., Shao, L. J., & Gong, B. (2021b). Pore characteristics and influencing factors of coal reservoir in Tucheng Mining Area of Guizhou Province. *Safety in Coal Mines, 52*(7), 21–26+32.

Tao, L., Banghua, Y., Yong, L., & Dengke, W. (2021). Grouting fractured coal permeability evolution based on industrial CT scanning. *Geofluids, 2021*, 1–12.

Topal, T., & Sözmen, B. (2003). Deterioration mechanisms of tuffs in Midas monument. *Engineering Geology, 68*(3–4), 201–223.

Tse, R., & Cruden, D. M. (1979). Estimating joint roughness coefficients. *International Journal of Rock Mechanics and Mining Sciences & Geomechanics Abstracts, 16*(5), 303–307.

Turturro, A. C., Caputo, M. C., & Gerke, H. H. (2022). Mercury intrusion porosimetry and centrifuge methods for extended-range retention curves of soil and porous rock samples. *Vadose Zone Journal, 21*(1), e20176.

Ukpai, S. N. (2021). Stability analyses of dams using multidisciplinary geoscience approach for water reservoir safety: Case of Mpu damsite. *Bulletin of Engineering Geology and the Environment, 80*(3), 2149–2170.

Verstrynge, E., Adriaens, R., Elsen, J., & Van Balen, K. (2014). Multi-scale analysis on the influence of moisture on the mechanical behavior of ferruginous sandstone. *Construction and Building Materials, 54*, 78–90.

Vervoort, A. (2021). Uplift of the surface of the earth above abandoned coal mines. Part A: Analysis of satellite data related to the movement of the surface. *International Journal of Rock Mechanics and Mining Sciences, 148*, 104896.

Vogler, D., Amann, F., Bayer, P., & Elsworth, D. (2016). Permeability evolution in natural fractures subject to cyclic loading and gouge formation. *Rock Mechanics and Rock Engineering, 49*(9), 3463–3479.

Wang L, Yu Q., 2016. The effect of moisture on the methane adsorption capacity of shales: a study case in the eastern Qaidam Basin in China. J Hydrol. 542, 487–505.

Wang, J., Li, J., & Xu, M. (2010). Development and application of simulation test system for water inrush from the water-conducting collapse column. *Journal of Mining & Safety Engineering, 27*(3), 305–309. (in Chinese).

Wang, J., Zhang, Q., Song, Z., Liu, X., Wang, X., & Zhang, Y. (2022). Microstructural variations and damage evolvement of salt rock under cyclic loading. International Journal of Rock Mechanics and Mining Sciences. 152, 105078.

Wang, L., Chen, Z., Kong, H., & Ni, X. (2014). An experimental study of the influence of seepage pressure and initial porosity on variable mass seepage for broken mudstone. *Journal of Mining & Safety Engineering, 31*(3), 462–475. (in Chinese).

Wang, F., Tu, S., Zhang, C., Zhang, Y., & Bai, Q. (2016a). Evolution mechanism of water-flowing zones and control technology for longwall mining in shallow coal seams beneath gully topography. *Environment and Earth Science, 75*(19), 1309.

Wang, M., Chen, Y., Ma, G., Zhou, J., & Zhou, C. (2016b). Influence of surface roughness on nonlinear flow behaviors in 3D self-affine rough fractures: Lattice Boltzmann simulations. *Advances in Water Resources, 96*, 373–388.

Wang, Z., Shen, L., Li, S., & Xu, Z. (2017). Seepage characteristics of a single fracture based on lattice Boltzmann method. *Rock and Soil Mechanics, 38*(4), 1203–1210. (in Chinese).

Wang, C., Zhang, C., Zhao, X., Liao, L., & Zhang, S. (2018). Dynamic structural evolution of overlying strata during shallow coal seam longwall mining. *International Journal of Rock Mechanics and Mining Sciences, 103*, 20–32.

Wang, F., Liang, N., & Li, G. (2019). Damage and failure evolution mechanism for coal pillar dams affected by water immersion in underground reservoirs. *Geofluids, 2019*, 2985691.

Wang, H., Shi, R., Deng, D., Jiang, Y., Wang, G., & Gong, W. (2020a). Characteristic of stress evolution on fault surface and coal bursts mechanism during the extraction of longwall face in Yima mining area, China. *Journal of Structural Geology, 136*, 104071.

Wang, J., Zhang, Y., Qin, Z., Song, S., & Lin, P. (2020b). Analysis method of water inrush for tunnels with damaged water-resisting rock mass based on finite element method-smooth particle hydrodynamics coupling. *Computers and Geotechnics, 126*, 103725.

Wang, L., Kong, H., & Karakus, M. (2020c). Hazard assessment of groundwater inrush in crushed rock mass: An experimental investigation of mass-loss-induced change of fluid flow behavior. *Engineering Geology, 277*, 105812.

Wang, G., Qin, X., Han, D., & Liu, Z. (2021a). Study on seepage and deformation characteristics of coal microstructure by 3D reconstruction of CT images at high temperatures. *International Journal of Mining Science and Technology, 31*(2), 175–185.

Wang, Y., Huang, W., Li, Z., Chang, L., Li, D., Chen, R., & Lv, B. (2021b). Small furnace for the small angle X-ray scattering (SAXS) and wide angle X-ray scattering (WAXS) characterization of the high temperature carbonization of coal. *Instrumentation Science and Technology, 49*(4), 445–456.

Wang, Y., Li, M., Chen, B., & Dai, H. (2015). Experimental study on ultrasonic wave characteristics of coal samples under dry and water saturated conditions. Journal of China Coal Society. (10), 2445–2450.

Wang, C., Zhang, C., Li, Z., & Zhou, J. (2023a). Using discrete element numerical simulation to determine effect of persistent fracture morphology on permeability stress sensitivity. *International Journal for Numerical and Analytical Methods in Geomechanics, 47*, 1–20.

Wang, X. L., Pan, J. N., Wang, K., Li, J. X., Cheng, N. N., & Li, M. (2023b). Characteristics of micro-CT scale pore-fracture of tectonic ally deformed coal and their controlling effect on performance. *Journal of China Coal Society, 48*(3), 1325–1334.

Whittles, D. N., Lowndes, I. S., Kingman, S. W., Yates, C., & Jobling, S. (2006). Influence of geotechnical factors on gas flow experienced in a UK longwall coal mine panel. *International Journal of Rock Mechanics and Mining Sciences, 43*(3), 369–387.

Wu, J., Han, G., Feng, M., Kong, H., Yu, B., Wang, L., & Gao, Y. (2019). Mass-loss effects on the flow behavior in broken argillaceous red sandstone with different particle-size distributions. *Comptes Rendus Mécanique, 347*(6), 504–523.

Wu, B., Liu, K., & Guo, D. (2020a). Study on the change law of mechanical properties of gritstone under the influence of mine water. *Journal Of Mining Science And Technology, 5*(6), 632–637. (in Chinese).

Wu, B., Wang, W., & Guo, D. (2020b). Strength damage and AE characteristics of fractured sandstone under the influence of water intrusion times. *Journal of Mining & Safety Engineering, 37*(5), 1054. (in Chinese).

Wu, Y., Wang, D., Wei, J., Yao, B., Zhang, H., Fu, J., & Zeng, F. (2022). Damage constitutive model of gas-bearing coal using industrial CT scanning technology. *Journal of Natural Gas Science and Engineering, 101*, 104543.

Xiao, W., Zhang, D., & Wang, X. (2020). Experimental study on progressive failure process and permeability characteristics of red sandstone under seepage pressure. *Engineering Geology, 265*, 105406.

Xie, K., Jiang, D., Sun, Z., Chen, J., Zhang, W., & Jiang, X. (2018a). NMR, MRI and AE statistical study of damage due to a low number of wetting–drying cycles in sandstone from the Three Gorges Reservoir area. *Rock Mechanics and Rock Engineering, 51*(11), 3625–3634.

Xie, P., Li, W., Yang, D., & Jiao, J. (2018b). Hydrogeological model for groundwater prediction in the Shennan mining area, China. *Mine Water and the Environment, 37*(3), 505–517.

Xiong, D., Zhao, Z., Su, C., & Wang, Y. (2011). Experimental study on effect of water saturated on mechanical properties of rock in coal measure strata. *Chinese Journal of Rock Mechanics and Engineering, 30*(5), 998–1006. (in Chinese).

Xu, B. H., & Yu, A. B. (1997). Numerical simulation of the gas-solid flow in a fluidized bed by combining discrete particle method with computational fluid dynamics. *Chemical Engineering Science, 52*(16), 2785–2809.

Xu, C., Li, X., Wang, K., Guo, C., Fu, Q., & Ju, Y. (2020). Reasonable coal pillar width of protective layer during high-gas coal seams mining. *Journal of China University of Mining and Technology, 49*(3), 445–452.

Xu, X., He, F., Li, X., & He, W. (2021). Research on mechanism and control of asymmetric deformation of gob side coal roadway with fully mechanized caving mining. *Engineering Failure Analysis, 120*, 105097.

Xu, F., Jin, D., Gao, Z., He, Y., Wang, S., Shi, L., et al. (2022). Study on law of groundwater resources leakage under high intensity repeated mining. *Coal Science and Technology, 50*(11), 131–139.

Yang, C., Mao, H., Wang, X., Li, X., & Chen, J. (2006). Study on variation of microstructure and mechanical properties of water-weakening slates. *Rock and Soil Mechanics, 27*(12), 2090–2098. (in Chinese).

Yang, W., Lin, B., Qu, Y., Li, Z., Zhai, C., Jia, L., & Zhao, W. (2011). Stress evolution with time and space during mining of a coal seam. *International Journal of Rock Mechanics and Mining Sciences, 48*(7), 1145–1152.

Yang, T., Liu, H. Y., & Tang, C. A. (2017). Scale effect in macroscopic permeability of jointed rock mass using a coupled stress–damage–flow method. *Engineering Geology, 228*, 121–136.

Yang, D., Qi, X., Chen, W., Wang, S., & Yang, J. (2018a). Anisotropic permeability of coal subjected to cyclic loading and unloading. *International Journal of Geomechanics, 18*(8), 04018093.

Yang, X., Wang, J., Hou, D., Zhu, C., & He, M. (2018b). Effect of dry-wet cycling on the mechanical properties of rocks: A laboratory-scale experimental study. *Processes, 6*(10), 199.

Yang, S. Q., Tian, W. L., Ranjith, P. G., Liu, X. R., Chen, M., & Cai, W. (2022). Three-dimensional failure behavior and cracking mechanism of rectangular solid sandstone containing a single fissure under triaxial compression. *Rock Mechanics Bulletin, 1*(1), 100008.

Yang, S., & Yu, Q. (2022). The role of fluid-rock interactions in permeability behavior of shale with different pore fluids. International Journal of Rock Mechanics and Mining Sciences, 150, 105023.

Yao, Y., Liu, D., Che, Y., Tang, D., Tang, S., & Huang, W. (2009). Non-destructive characterization of coal samples from China using microfocus X-ray computed tomography. *International Journal of Coal Geology, 80*(2), 113–123.

Yao, Q., Chen, T., Ju, M., Liang, S., Liu, Y., & Li, X. (2016). Effects of water intrusion on mechanical properties of and crack propagation in coal. *Rock Mechanics and Rock Engineering, 49*(12), 4699–4709.

Yao, B., Wang, L., Wei, J., Li, Z., & Liu, X. (2018). A deformation seepage-erosion coupling model for water outburst of Karst collapse pillar and its application. *Journal of China Coal Society, 43*(7), 2007–2013. (in Chinese).

Yao, Q., Chen, T., Tang, C., Sedighi, M., Wang, S., & Huang, Q. (2019a). Influence of moisture on crack propagation in coal and its failure modes. *Engineering Geology, 258*, 105156.

Yao, Q., Hao, Q., Chen, X., Zhou, B., & Fang, J. (2019b). Design on the width of coal pillar dam in coal mine groundwater reservoir. *Journal of China Coal Society, 44*(3), 231–239.

Yao, Q., Tang, C., Xia, Z., Liu, X., Zhu, L., Chong, Z., & Hui, X. (2020a). Mechanisms of failure in coal samples from underground water reservoir. *Engineering Geology, 267*, 105494.

Yao, Q., Wang, W., Zhu, L., Xia, Z., Tang, C., & Wang, X. (2020b). Effects of moisture conditions on mechanical properties and AE and IR characteristics in coal–rock combinations. *Arabian Journal of Geosciences, 13*(14), 1–15.

Yao, W., Li, C., Zhan, H., Zhou, J. Q., Criss, R. E., Xiong, S., & Jiang, X. (2020c). Multiscale study of physical and mechanical properties of sandstone in Three Gorges Reservoir region subjected to cyclic wetting–drying of yangtze river water. *Rock Mechanics and Rock Engineering, 53*(5), 2215–2231.

Yu, B., Chen, Z., & Wu, J. (2018). Experimental investigation on seepage stability of filling material of karst collapse pillar in mining engineering. *Advances in Civil Engineering, 2018*, 3986490.

Yu, C., Tang, S., Duan, D., Zhang, Y., Liang, Z., Ma, K., & Ma, T. (2019). The effect of water on the creep behavior of red sandstone. *Engineering Geology, 253*, 64–74.

Yu, C., Tian, W., Zhang, C., Chai, S., Cheng, X., & Wang, X. (2021). Temperature-dependent mechanical properties and crack propagation modes of 3D printed sandstones. *International Journal of Rock Mechanics and Mining Sciences, 146*, 104868.

Yu, L., Yao, Q., Chong, Z., Li, Y., Xu, Q., Xie, H., & Ye, P. (2022). Mechanical and micro-structural damage mechanisms of coal samples treated with dry–wet cycles. *Engineering Geology, 304*, 106637.

Yuan, K., Liao, L., Wan, H., Wang, C., & Wang, C. (2011). Quantitative analysis of cristobalite and α-quartz in bentonite by X-ray powder diffraction—Comparison between external standard and K-value method. *Journal of the Chinese Ceramic Society, 39*(2), 377–382.

Yuan, L., Wang, E. Y., Ma, Y. K., Liu, Y. B., & Li, X. L. (2023). Research progress of coal and rock dynamic disasters and scientific and technological problems in China. *Journal of China Coal Society, 48*, 1–37.

Zang, A., Wagner, C. F., & Dresen, G. (1996). Acoustic emission, microstructure, and damage model of dry and wet sandstone stressed to failure. Journal of Geophysical Research: Solid Earth, 101(B8), 17507–17521.

Zhang, Y. (2021). Mechanism of water inrush of a deep mining floor based on coupled mining pressure and confined pressure. *Mine Water and the Environment, 40*(2), 366–377.

Zhang, L., & Einstein, H. H. (2004). Using RQD to estimate the deformation modulus of rock masses. *International Journal of Rock Mechanics and Mining Sciences, 41*(2), 337–341.

Zhang, Z., Jiang, Q., Zhou, C., & Liu, X. (2014). Strength and failure characteristics of Jurassic Red-Bed sandstone under cyclic wetting–drying conditions. *Geophysical Journal International, 198*(2), 1034–1044.

Zhang, B., Bai, H., & Zhang, K. (2016a). Experimental research on seepage mutation mechanism of collapse column medium. *Rock and Soil Mechanics, 37*(3), 745–752. (in Chinese).

Zhang, C., Tu, S., Zhang, L., Wang, F., Bai, Q., & Tu, H. (2016c). The numerical simulation of permeability rules in protective seam mining. *International Journal of Oil, Gas and Coal Technology, 13*(3), 243–259.

Zhang, J. F., Ye, J. B., Chen, J. S., & Li, S. L. (2016d). A preliminary study of measurement and evaluation of breakstone grain shape. *Rock and Soil Mechanics, 37*(2), 343–349.

Zhang, Z., Zhang, R., Xie, H., Gao, M., & Xie, J. (2016e). Mining-induced coal permeability change under different mining layouts. *Rock Mechanics and Rock Engineering, 49*, 3753–3768.

Zhang, F., Dontsov, E., & Mack, M. (2017a). Fully coupled simulation of a hydraulic fracture interacting with natural fractures with a hybrid discrete-continuum method. *International Journal for Numerical and Analytical Methods in Geomechanics, 41*(13), 1430–1452.

Zhang, G., He, F., Jia, H., & Lai, Y. (2017b). Analysis of gateroad stability in relation to yield pillar size: A case study. *Rock Mechanics and Rock Engineering, 50*(5), 1263–1278.

Zhang, C., Zhang, L., Tu, S., Hao, D., & Teng, T. (2018a). Experimental and numerical study of the influence of gas pressure on gas permeability in pressure relief gas drainage. *Transport in Porous Media, 124*, 995–1015.

Zhang, C., Zhang, L., Zhao, Y., & Wang, W. (2018b). Experimental study of stress–permeability behavior of single persistent fractured coal samples in the fractured zone. *Journal of Geophysics and Engineering, 15*(5), 2159–2170.

Zhang, N., Wang, S., Yan, C., Gao, J., Guo, R., Wang, H. (2019a). Pore structure evolution of hydration damage of mudstone based on NMR technology. *J. China Coal Soc. 44*(S1), 110–117.

Zhang, C., Liu, J., Zhao, Y., Zhang, L., & Guo, J. (2019b). A fluid-solid coupling method for the simulation of gas transport in porous coal and rock media. *Energy Science & Engineering, 7*(5), 1913–1924.

Zhang, C., Tu, S., & Zhao, Y. (2019c). Compaction characteristics of the caving zone in a longwall goaf: A review. *Environmental Earth Sciences, 78*(1), 27–46.

Zhang, C., Zhang, L., & Wang, W. (2019d). The axial and radial permeability testing of coal under cyclic loading and unloading. *Arabian Journal of Geosciences, 12*(11), 1–19.

Zhang, F., Damjanac, B., & Maxwell, S. (2019e). Investigating hydraulic fracturing complexity in naturally fractured rock masses using fully coupled multiscale numerical modeling. *Rock Mechanics and Rock Engineering, 52*(12), 5137–5160.

Zhang, N., Wang, S., Yan, C., Gao, J., Guo, R., & Wang, H. (2019f). Pore structure evolution of hydration damage of mudstone based on NMR technology. *Journal of China Coal Society, 44*(S1), 110–117. (in Chinese).

Zhang, Q., Huang, X., Zhu, H., & Li, J. (2019g). Quantitative assessments of the correlations between rock mass rating (RMR) and geological strength index (GSI). *Tunnelling and Underground Space Technology, 83*, 73–81.

Zhang, T., Pang, M., Peng, W., Liu, N., & Huang, Y. (2019h). Seepage stability of cemented and fractured coal rock mass under tri-axial stress. *Journal of Mining & Safety Engineering, 36*(4), 834–847. (in Chinese).

Zhang, T., Gan, Q., Zhao, Y., Zhu, G., Nie, X., Yang, K., & Li, J. (2019i). Investigations into mining-induced stress–fracture–seepage field coupling effect considering the response of key stratum and composite aquifer. *Rock Mechanics and Rock Engineering, 52*(10), 4017–4031.

Zhang, C., Bai, Q., & Chen, Y. (2020a). Using stress path-dependent permeability law to evaluate permeability enhancement and coalbed methane flow in protected coal seam: A case study. *Geomechanics and Geophysics for Geo-Energy and Geo-Resources, 6*(3), 1–25.

Zhang, C., Tu, S., & Zhang, L. (2020b). Field measurements of compaction seepage characteristics in longwall mining goaf. *Natural Resources Research, 29*(2), 905–917.

Zhang, F., Wang, T., Liu, F., Peng, M., Furtney, J., & Zhang, L. (2020c). Modeling of fluid-particle interaction by coupling the discrete element method with a dynamic fluid mesh: Implications to suffusion in gap-graded soils. *Computers and Geotechnics, 124*, 103617.

Zhang, C., Han, P., Wang, F., & He, X. (2021a). Study on the stability of residual coal pillar in underground reservoir under the effect of mining and water invasion. *Journal of China University of Mining and Technology, 50*(2), 220–227. (in Chinese).

Zhang, C., Jia, S., Bai, Q., Zhang, H., Chen, Y., & Jiao, Y. (2021b). CFD-DEM coupled simulation of broken rock mass movement during water seepage in an underground goaf reservoir. *Mine Water and the Environment, 40*(4), 1048–1060.

Zhang, C., Wang, F., & Bai, Q. (2021c). Underground space utilization of coalmines in China: A review of underground water reservoir construction. *Tunnelling and Underground Space Technology, 107*, 103657.

Zhang, T., Zhang, X., Pang, M., Liu, N., Zhang, S., & Gao, H. (2021e). Effect of particle loss on the pore structure and emergent behavior of karst column fills. *Journal of China Coal Society, 46*(10), 3245–3254. (in Chinese).

Zhang, W., Zhang, D., & Zhao, J. (2021f). Experimental investigation of water sensitivity effects on microscale mechanical behavior of shale. *International Journal of Rock Mechanics and Mining Sciences, 145*, 104837.

Zhang, Y., Shi, F., Wang, J., Zhang, J., Lu, H., & Sun, X. (2021g). Study on softening characteristics of mudstones in shallow buried strata in Xinjiang and countermeasures for tunnel support. *Journal Of Mining Science And Technology, 6*(1), 42–50.

Zhang, C., Li, B., Song, Z., Liu, J., & Zhou, J. (2022a). Breakage mechanism and pore evolution characteristics of gangue materials under compression. *Acta Geotechnica, 17*(11), 4823–4835.

Zhang, C., Zhao, Y., & Bai, Q. (2022b). 3D DEM method for compaction and breakage characteristics simulation of broken rock mass in goaf. *Acta Geotechnica, 17*(7), 2765–2781.

Zhang, C., Zhao, Y., Han, P., & Bai, Q. (2022c). Coal pillar failure analysis and instability evaluation methods: A short review and prospect. *Engineering Failure Analysis, 138*, 106344.

Zhang, L., Kan, Z., Zhang, C., & Tang, J. (2022e). Experimental study of coal flow characteristics under mining disturbance in China. *International Journal of Coal Science & Technology, 9*(1), 1–16.

Zhang, S., Zou, C., Peng, C., Yan, L., & Wu, X. (2022f). Pore structure and its effect on acoustic velocity and permeability of reef-shoal carbonates in the Tarim Basin, Northwest China. *Journal of Geophysics and Engineering, 19*(6), 1340–1354.

Zhang, T., Zhang, C., & Song, Z. (2022g). Efficient and simple simulation method for matching stress-permeability test results. *Energy Science & Engineering, 10*(4), 1214–1226.

Zhang, C., Bai, Q., & Han, P. (2023a). A review of water rock interaction in underground coal mining: Problems and analysis. *Bulletin of Engineering Geology and the Environment, 82*(5), 157.

Zhang, C., Bai, Q., Han, P., Wang, L., Wang, X., & Wang, F. (2023b). Strength weakening and its micromechanism in water–rock interaction, a short review in laboratory tests. *International Journal of Coal Science & Technology, 10*(1), 10.

Zhang, C., Jia, S., Wang, F. T., Liu, J. B., Wang, W., & Qiao, Y. D. (2023c). Evolution characteristics of pores and fracture in coal body under water-rock action and its experimental research on driving mechanism. *Journal of Basic Science and Engineering, 31*(1), 185–196.

Zhao, J., & Konietzky, H. (2020). Numerical analysis and prediction of ground surface movement induced by coal mining and subsequent groundwater flooding. *International Journal of Coal Geology, 229*, 103565.

Zhao, Y., Liu, S., Elsworth, D., Jiang, Y., & Zhu, J. (2014). Pore structure characterization of coal by synchrotron small-angle X-ray scattering and transmission electron microscopy. *Energy & Fuels, 28*(6), 3704–3711.

Zhao, Y., Liu, S., Jiang, Y., Wang, K., & Huang, Y. (2016). Dynamic tensile strength of coal under dry and saturated conditions. Rock Mechanics and Rock Engineering, 49(5), 1709–1720.

Zhao, Y., Xue, S., Han, S., Chen, Z., Liu, S., Elsworth, D., & Chen, D. (2017a). Effects of microstructure on water imbibition in sandstones using X-ray computed tomography and neutron radiography. *Journal of Geophysical Research: Solid Earth, 122*(7), 4963–4981.

Zhao, Z., Yang, J., Zhang, D., & Peng, H. (2017b). Effects of wetting and cyclic wetting–drying on tensile strength of sandstone with a low clay mineral content. *Rock Mechanics and Rock Engineering, 50*(2), 485–491.

Zhao, Z., Yang, J., Zhou, D., & Chen, Y. (2017c). Experimental investigation on the wetting-induced weakening of sandstone joints. *Engineering Geology, 225*, 61–67.

Zhao, Y., Sun, Y., Liu, S., Chen, Z., & Yuan, L. (2018). Pore structure characterization of coal by synchrotron radiation nano-CT. *Fuel, 215*, 102–110.

Zhao, Y., Ren, S., Jiang, D., Liu, R., Wu, J., & Jiang, X. (2018a). Influence of wetting-drying cycles on the pore structure and mechanical properties of mudstone from Simian Mountain. *Construction and Building Materials, 191*, 923–931.

Zhao, Y., Xue, S., Han, S., He, L., & Chen, Z. (2018c). Characterization of unsaturated diffusivity of tight sandstones using neutron radiography. *International Journal of Heat and Mass Transfer, 124*, 693–705.

Zhao, Z., Guo, T., Ning, Z., Dou, Z., Dai, F., & Yang, Q. (2018d). Numerical modeling of stability of fractured reservoir bank slopes subjected to water–rock interactions. *Rock Mechanics and Rock Engineering, 51*(8), 2517–2531.

Zhao, L., Sun, C., Yan, P., Zhang, Q., Wang, S., Luo, S., & Mao, Y. (2019a). Dynamic changes of nitrogen and dissolved organic matter during the transport of mine water in a coal mine underground reservoir: Column experiments. *Journal of Contaminant Hydrology, 223*, 103473.

Zhao, Y., Peng, L., Liu, S., Cao, B., Sun, Y., & Hou, B. (2019b). Pore structure characterization of shales using synchrotron SAXS and NMR cryoporometry. *Marine and Petroleum Geology, 102*, 116–125.

Zhao, Y., Song, H., Liu, S., Zhang, C., Dou, L., & Cao, A. (2019c). Mechanical anisotropy of coal with considerations of realistic microstructures and external loading directions. *International Journal of Rock Mechanics and Mining Sciences, 116*, 111–121.

Zhao, Y., Wu, Y., Han, S., Xue, S., Fan, G., Chen, Z., & El Abd, A. (2019d). Water sorptivity of unsaturated fractured sandstone: Fractal modeling and neutron radiography experiment. *Advances in Water Resources, 130*, 172–183.

Zhao, C., Dong, S., Wang, H., Jin, D., Li, C., Wang, S., & Liu, Y. (2020a). Analysis of water inrush from boreholes for drainage of confined aquifer by upward boreholes in underground coal mining face. *Journal of China Coal Society, 45*(S1), 405–414.

Zhao, Y., Wang, C., & Bi, J. (2020c). Analysis of fractured rock permeability evolution under unloading conditions by the model of elastoplastic contact between rough surfaces. *Rock Mechanics and Rock Engineering, 53*(12), 5795–5808.

Zhao, B., Li, Y., Huang, W., Yang, J., Sun, J., Li, W., & Zhang, L. (2021a). Mechanical characteristics of red sandstone under cyclic wetting and drying. *Environment and Earth Science, 80*(22), 1–12.

Zhao, J., Konietzky, H., Herbst, M., & Morgenstern, R. (2021b). Numerical simulation of flooding induced uplift for abandoned coal mines: Simulation schemes and parameter sensitivity. *International Journal of Coal Science & Technology, 8*(6), 1238–1249.

Zhong, C., Zhang, Z., Ranjith, P. G., Lu, Y., & Choi, X. (2019). The role of pore water plays in coal under uniaxial cyclic loading. *Engineering Geology, 257*, 105125.

Zhou, Z. Y., Kuang, S. B., Chu, K. W., & Yu, A. (2010). Discrete particle simulation of particle–fluid flow: Model formulations and their applicability. *Journal of Fluid Mechanics, 661*, 482–510.

Zhou, J. Q., Hu, S. H., Fang, S., Chen, Y. F., & Zhou, C. B. (2015). Nonlinear flow behavior at low Reynolds numbers through rough-walled fractures subjected to normal compressive loading. *International Journal of Rock Mechanics and Mining Sciences, 80*, 202–218.

Zhou, J., Hu, S., Chen, Y., Wang, M., & Zhou, C. (2016). The friction factor in the Forchheimer equation for rock fractures. *Rock Mechanics and Rock Engineering, 49*(8), 3055–3068.

Zhou, Z., Cai, X., Chen, L., Cao, W., Zhao, Y., & Xiong, C. (2017). Influence of cyclic wetting and drying on physical and dynamic compressive properties of sandstone. *Engineering Geology, 220*, 1–12.

Zhou, H. W., Zhong, J. C., Ren, W. G., Wang, X. Y., & Yi, H. Y. (2018a). Characterization of pore-fracture networks and their evolution at various measurement scales in coal samples using X-ray μCT and a fractal method. *International Journal of Coal Geology, 189*, 35–49.

Zhou, Q., Herrera, J., & Hidalgo, A. (2018b). The numerical analysis of fault-induced mine water inrush using the extended finite element method and fracture mechanics. *Mine Water and the Environment, 37*(1), 185–195.

Zhou, Z., Cai, X., Ma, D., Cao, W., Chen, L., & Zhou, J. (2018c). Effects of water content on fracture and mechanical behavior of sandstone with a low clay mineral content. *Engineering Fracture Mechanics, 193*, 47–65.

Zhou, Z., Cai, X., Ma, D., Chen, L., Wang, S., & Tan, L. (2018d). Dynamic tensile properties of sandstone subjected to wetting and drying cycles. *Construction and Building Materials, 182*, 215–232.

Zhou, Z., Xiong, C., Cai, X., Zhao, Y., Li, X., & Du, K. (2018e). Mechanical and infrared radiation properties of sandstone with different water contents under uniaxial compression. *Journal of Central South University (Science and Technology), 49*(5), 1189–1196.

Zhu, J., Zhang, M., Chuan, L. J., Tang, J., & Zhao, F. (2016). Experimental study on coal strain induced by methane sorption/desorption and effect of pore features. *Chinese Journal of Rock Mechanics and Engineering, 35*(S1), 2620–2626.

Zhu, D., Tu, S., Ma, H., Wei, H., Li, H., & Wang, C. (2019). Modeling and calculating for the compaction characteristics of waste rock masses. *International Journal for Numerical and Analytical Methods in Geomechanics, 43*(1), 257–271.

Zhu, D., Wu, Y., Liu, Z., Dong, X., & Yu, J. (2020a). Failure mechanism and safety control strategy for laminated roof of wide-span roadway. *Engineering Failure Analysis, 111*, 104489.

Zhu, J., Deng, J., Chen, F., Huang, Y., & Yu, Z. (2020b). Water saturation effects on mechanical and fracture behavior of marble. *International Journal of Geomechanics, 20*(10), 04020191.

Zhu, J., Deng, J., Chen, F., Ma, Y., & Yao, Y. (2021). Water-weakening effects on the strength of hard rocks at different loading rates: An experimental study. *Rock Mechanics and Rock Engineering, 54*(8), 4347–4353.

Zhu, L. P., Shen, W. Q., He, M. C., & Shao, J. F. (2022). Contribution of atomistic study to better understand water saturation effect on mechanical behavior of clayey rocks in triaxial compression. Computers and Geotechnics, 146, 104738.

Zou, L., Jing, L., & Cvetkovic, V. (2017). Shear-enhanced nonlinear flow in rough-walled rock fractures. *International Journal of Rock Mechanics and Mining Sciences, 97*, 33–45.

GPSR Compliance

*The European Union's (EU) General Product Safety Regulation (GPSR) is a set of rules that requires consumer products to be safe and our obligations to ensure this.*

*If you have any concerns about our products, you can contact us on ProductSafety@springernature.com*

In case Publisher is established outside the EU, the EU authorized representative is:

Springer Nature Customer Service Center GmbH
Europaplatz 3
69115 Heidelberg, Germany

**Batch number: 10425614**

Printed by Printforce, the Netherlands